Materials Recycling Handbook

Materials Recycling Handbook

Eden Sparks

MURPHY & MOORE
www.murphy-moorepublishing.com

Published by Murphy & Moore Publishing,
1 Rockefeller Plaza,
New York City, NY 10020, USA

ISBN: 978-1-63987-352-4

Cataloging-in-Publication Data

Materials recycling handbook / Eden Sparks.
p. cm.
Includes bibliographical references and index.
ISBN 978-1-63987-352-4
1. Recycling (Waste, etc.). 2. Waste products. 3. Recycling industry. 4. Refuse and refuse disposal.
5. Factory and trade waste. I. Sparks, Eden.
TD794.5 .M38 2022
363.728 2--dc23

For information on all Murphy & Moore Publications
visit our website at www.murphy-moorepublishing.com

MURPHY & MOORE

Table of Contents

Preface   VII

Chapter 1 **Introduction to Recycling**   1
- Recyclate   7
- Common Types of Recycling   9
- Economic aspects of Recycling   13
- Scrap   30
- Sustainable Solid Waste Recycling   34
- Downcycling   39
- Upcycling   42
- Waste Hierarchy   48
- Waste Minimisation   51
- Recursive Recycling   57
- Recommerce   58
- Recycling Symbol   62
- Recycling Codes   65

Chapter 2 **Recycling of Materials**   71
- Ferrous Metals Recycling   71
- Non-Ferrous Recycling   73
- Plastic Recycling   78
- Aluminium Recycling   88
- Concrete Recycling   92
- Cotton Recycling   95
- Copper Recycling   96
- Glass Recycling   100
- Gypsum Recycling   105
- Paper Recycling   107
- Timber Recycling   111

- Paint Recycling   117
- Vegetable Oil Recycling   121
- Water Recycling   122

Chapter 3 Product Recycling   **128**
- Cardboard Recycling   128
- Automotive Oil Recycling   131
- Oil Filters Recycling   139
- Appliance Recycling   141
- Battery Recycling   144
- Fluorescent Lamp Recycling   156
- Textile Recycling   158
- Tire Recycling   163
- Bottle Recycling   170
- Drug Recycling   171
- Recycling Processes for Photovoltaic Modules   178

Chapter 4 E-waste Recycling   **188**
- Electronic Waste   188
- E-waste Recycling   191
- Generation, Composition, Collection, Treatment and
  Disposal System   198

Chapter 5 Vehicle Recycling   **209**
- Car Recycling   212
- Ship Recycling   218
- Green Ship Recycling   225

Permissions

Index

Preface

It is with great pleasure that I present this book. It has been carefully written after numerous discussions with my peers and other practitioners of the field. I would like to take this opportunity to thank my family and friends who have been extremely supporting at every step in my life.

The process of conversion of waste materials into new ones is known as recycling. Materials recycling are used to reduce the usage of fresh raw materials and lower greenhouse gas emissions. It is considered an alternative to conventional disposal of waste. It is an important component in modern waste reduction. It is used to recycle a variety of materials such as metal, plastic, cardboard, batteries, electronics and paper. Depending upon the material, different techniques are used for its recycling. The field is mainly helpful in producing a fresh supply of the same material that is recycled. Materials recycling is an upcoming field of science that has undergone rapid development over the past few decades. The book unfolds the innovative aspects of materials recycling which will be crucial for the understanding of the subject matter. It aims to serve as a resource guide for students and experts alike and contribute to the growth of the discipline.

The chapters below are organized to facilitate a comprehensive understanding of the subject:

Chapter – Introduction to Recycling

Recycling is an alternative to conventional waste disposal process that converts waste materials into new materials and objects. It prevents the waste of useful materials and consumption of new raw materials and further reduces energy usage, and water and air pollution. This chapter delves into the subject of recycling for a thorough understanding of it.

Chapter – Recycling of Materials

Recycling can be carried out on a variety of materials such as concrete, glass, plastic, timber, cotton, gypsum, paper, paint, vegetable, timber and metals including aluminium, copper, iron and steel, etc. The topics elaborated in this chapter will help in gaining a better perspective of the recycling of different materials.

Chapter – Product Recycling

Products made from a variety of materials can be recycled using a number of processes. It includes products and items like cardboard, tire, bottle, lamps, textiles, batteries, oil filters, etc. This chapter closely examines product recycling to provide an extensive understanding of the subject.

Chapter – E-waste Recycling

E-waste refers to the discarded electronic devices. It includes electronics wastes such as computer monitors, printers, scanners, keyboards, circuit boards, clocks, flashlight, calculators, phones, etc. E-waste recycling is the reuse and reprocessing of electronics and electricals. This chapter discusses the subject of E-waste recycling in detail.

Chapter – Vehicle Recycling

Vehicles, at the end of their useful life, are dismantled for their spare parts and the process is termed as vehicle recycling. Different kinds of vehicles can be recycled such as cars, ships, trucks, etc. This chapter closely examines vehicle recycling to provide an extensive understanding of the subject.

Eden Sparks

1

Introduction to Recycling

Recycling is an alternative to conventional waste disposal process that converts waste materials into new materials and objects. It prevents the waste of useful materials and consumption of new raw materials and further reduces energy usage, and water and air pollution. This chapter delves into the subject of recycling for a thorough understanding of it.

Recycling is the process of breaking down and re-using materials that would otherwise be thrown away as trash. Many communities and businesses make it easy to recycle by placing labeled containers in the open for public use, or providing bins for home and business owners who have curbside pickup.

There are numerous benefits to recycling, and with so many new technologies making even more materials recyclable, with everyone's help we can clean up our Earth. Recycling not only benefits the environment but also have a positive effect on the economy. Recycling is reported throughout human history but has come a long way since the time of Plato when humans re-used broken tools and pottery when materials were scarce. Today, there is a multitude of benefits that come from recycling as well as tons of items that can be recycled.

Importance of Recycling

The benefits are far reaching and everybody gains when people adopt recycling as an everyday habit. Whether it is a community effort to help beautify a dirty neighborhood street, or on a larger scale to help a business save hundreds to thousands of dollars on waste management, the advantages of well-maintained recycling program are endless.

Environmental Benefits

- By recycling people can prevent millions of tons of material from entering landfills saving space for garbage that cannot be re-purposed. Landfills not only pollute the environment but also hampers the beauty of the city.

- The pollutants that are released into the air and water can be greatly reduced with an increase in recycling.

- Greatly reduces the amount of energy used daily by not needing to produce new materials. In short, recycling reduces the greenhouse gas emissions into the atmosphere.

- If for absolutely nothing else, recycling keeps litter overflow to a minimum keeping the Earth looking beautiful.

- In terms of energy, a single light bulb can be powered for up to four hours with the energy saved from one recycled glass bottle.

- Conserves the Earth's natural resources like raw materials, minerals, trees, etc.

Economic Benefits

- Properly run recycling programs cost the government, taxpayers, and business owners less money than waste programs.

- Studies show that by continuing to increase positive recycling habits, the United States can create over one million jobs annually.

- People can even make money by collecting approved materials to a nearby recycling facility that will pay for the product.

- For every one job created in the waste management industry recycling creates four.

Common Recyclable Items

There are so many materials that can be recycled in today's society:

Metal

Metals that we use in our everyday life are often times recyclable. Being a very versatile material, recycling metal takes more than seventy percent less energy than it does to produce a completely new item:

- Aluminum foil: (As well as bakeware) can easily be recycled. By melting down the foil products and simply repurposing the metal aluminum can be recycled almost infinitely.

- Aluminum cans: Studies show that Americans drink at least one canned beverage per day while only recycling a little over forty-five percent. It would save immense amounts of energy to recycle and reuse them as opposed to making new ones.

- Steel and tin cans: Things like coffee cans, soup containers, vegetable cans, etc. are one of the most recycled materials in America. This is a comforting statistic considering on average about 100 million are used daily in the States.

Paper and Cardboard

Most people can look around themselves at just about any point in the day and see paper or paper products. Paper is a material that has no limits in the recycling world, and American are doing a great job recycling. Studies show that people are recycling about 334 pounds of paper annually. Paper and cardboard materials that can be recycled are:

- Corrugated cardboard: This makes up most of the cardboard in people's everyday lives. Over seventy percent of shipping boxes already having been repurposed from sawdust, woodchips, or other paper products. Other items recycled cardboard is used to make are things like cereal boxes, tissue paper, printing paper, and poster board.

- Magazines and newspaper: Many people still get magazines and newspapers on their mailboxes and on their front porches. Too frequently these are junk ads or unwanted publications that go directly into the trash. One ton of recycled paper can save enough energy to power an American household for over five months.

- Office paper and poster board: Most people interact with at least one piece of paper a day. Papers are in the mailbox, printer, briefcase next to the door, everywhere. Paper can be easily repurposed saving high production cost and energy levels for new products.

Glass

Glass bottles and jars are not quite as versatile as paper or metal products when it comes to recycling. Due to the various colors of glass, many items can only be repurposed into another of the same item. The different types of glass recycling typically pertain to the color of the bottle or jar:

- Flint glass: This is the term used to refer to clear glass items which make up the largest part of the glass market at just over sixty percent. Usually, items bottled in clear glass containers is not light sensitive and people want to have seen.

- Amber glass: One of the reasons glass can be hard to recycle is due to the fact that the colors cannot be removed. For instance amber, or brown, glass makes up less of the glass market than flint at thirty-one percent partially because it can only be made into other amber colored glass products when recycled. Generally, items that are sensitive to sunlight are stored in brown colored glass bottles and jars.

- Emerald glass: When one thinks emerald, or green, glass wine and beer bottles typically come to mind. This is because the items inside are sensitive to sunlight and temperature, however not quite as sensitive as products that need to be stored in brown glass containers.

Practicing good recycling habits is not difficult, nor is there a secret to being a "good recycler." There are lots of ways to help create and maintain a better community through recycling.

Ways to Help in Recycling

- Get involved with local recycling programs.

- Volunteer to educate a class at a local elementary school. Kids can be huge advocates for the war on trash easily holding everyone around them accountable for the things they throw in the garbage.

- Take some time to go around and pick up trash in your own neighborhood or surround ones.

- If your job does not have a receptacle for recyclable products, ask if they wouldn't mind providing one, or if they mind you providing one yourself.

- Make sure to practice proper recycling habits in your own life, nothing works better than leading by example.

For people interested in playing a bigger role in the world of recycling around them Google (or your favorite search engine) is a good place to start. With a little bit of research, anyone can become and expert and start implementing positive recycling habits in their own lives, as well as those of the people around them. One person can make a gigantic difference, and together we can all change the world.

Benefits of Recycling

Reduce the Size of Landfills: One of the biggest reasons why recycling has been promoted is that it does reduce the strain on our environment. By utilizing waste products in a constructive way, we can slowly decrease the size of our landfills. As the population grows, it will become difficult for the landfills to hold so much and trash. When this happens, our cities and beautiful landscapes will face pollution, poisoning and many health problems. The benefits of recycling are that it helps to keep the pollution in check and decrease it little by little.

Conserve Natural Resources: Scrap cars, old bottles, junk mail and used rubber tyres are becoming common features of our landfills. All of these may seem endless, but the resources required to make them are finishing off quickly. Recycling allows all of these junk items to be used over and over again so that new resources do not have to be exploited. It conserves natural resources such as water, minerals, coal, oil, gas and timber. Another one of the benefits of recycling is that it allows more emphasis to be put on creating technology to utilize what already receive large quantities of recyclable exists. This is why a number of industries support programs where they can material to convert into new items.

More Employment Opportunities: While you may feel that recycling is each person for himself, in reality it is a huge industry within itself. After you do the basic sorting out and deposit your trash for recycling, it has to be sorted and shipped off to the right places. This is done by thousands of workers, who are newly employed by the growing industry. Certainly, one of the major benefits of recycling is that it creates more jobs in the community and provides stability to the entire process. Throwing the trash away creates some six to seven jobs at best, where recycling can help create close to thirty jobs.

Offers Cash Benefits: Recycling is not all about being charitable and doing what is good for the environment. If it were so, everybody would recycle out of the goodness of their hearts. Most governments have policies in place which give financial benefits to those who recycle. People that take the aluminum cans or glass bottles to the recycling plant, get a cash benefit in return. In fact, many teenagers can pick up recycling as a way to make extra money on the side. Old newspapers, appliances, plastic, rubber, steel, copper and even beer cans can be sold for money.

Saves Money: An unexpected place where the benefits of recycling can be seen is our economy. A strong economy is one that is efficient in nature. What drags it down is having to pay for resources that are growing scarce in the country. Every bit of recycling counts when the economy does not have to pay for planting more forests, mining iron ore or purchasing fossil fuels from other countries. When the jobs increase, the economy gets a boost. As the cost of maintaining the current waste disposal system go down, all the money saved is diverted to where it is need the most.

Reduce Greenhouse Gas Emissions: When you recycle products, you tend to save energy which results in less greenhouse gas emissions. Greenhouse gases are primarily responsible for increase in global warming. It helps to reduce air and water pollution by cutting down the number of pollutants that are released into the environment. A recycling rate of 30% can is almost equivalent of removing 30 million cars from the roads.

Saves Energy: When you recycle aluminum cans, you can save 95% of the energy required to produce those cans from raw materials, energy saved from recycling one glass bottle is enough to light a light bulb for four hours. This clearly shows how much energy can be saved if recycling is taken on a larger scale. With this, the reliance on foreign oil is reduced which also helps you to save money in long run.

Stimulate the Use of Greener Technologies: With use of more recycling products, it has pushed people towards more greener technologies. Use of renewable energy sources like solar, wind, geothermal is on rise which has helped to conserve energy and reduce pollution.

Bring Different Groups and Communities Together: At the end of the day, recycling is an act that can bring a community together. Whether it is by picking up trash from the roads or collecting waste materials to raise money for schools and colleges. Many simple programs that make a community stronger can be built upon the many benefits of recycling. Many people have found that their collective efforts in proper waste disposal have made their towns cleaner and happier. Others have found friends and supporters in their mission to change the world. In fact, it is one the best ways to teach children about responsibility and taking an initiative.

Prevents Loss of Biodiversity: Less raw material is needed when you engage yourself in recycling products. The beauty of recycling is that it will help you to conserve resources and prevents loss of biodiversity, ecosystems and rainforests. Mining activities will reduce which is considered as dangerous for mine workers. Soil erosion and water pollution will be reduced which in turn will protect native plants and animals to survive in forests. Deforestation, which is on rise these days will significantly reduce if recycling is considered seriously by majority of people.

The benefits of recycling are simple, but the effect they can have are large. Which is why so many countries support the process and make sure that their citizens face no trouble at all when they want to take up recycling.

RECYCLATE

Materials that are recyclable are called recyclate. Whether a recyclate is suitable to fulfill the requirements of a new application is evaluated through mechanical (e.g. tensile strength, impact strength), physical (e.g. hydrolytic stability, gloss, surface roughness) and processing tests (e.g. extrusion, injection molding) in standardized conditions. Standard test methods to analyze the stability of polymers cover multiple extrusion, accelerated heat aging and UV light exposure.

Multiple extrusion evaluates the melt processing stability of a polymer or polymer formulation as high temperature, shear forces and oxygen in the polymer transformation process cause degradation. The melt properties are usually characterized by melt flow rate (MFR) or melt volume rate (MVR) according to ISO 1133. For example the processing stability of polypropylene (PP) was studied by multiple extrusion with the use of a twin screw extruder. Molecular weight distribution and polydispersity of PP after processing were found to be in good relation to rheological changes and the mechanisms of degradation of PP could be identified.

Accelerated heat aging is simulating long-term thermal stability of plastics and effectiveness of antioxidants. Samples are oven-aged at defined temperatures and conditions, e.g. according to ISO 4577, and the change of the properties, e.g. visual appearance, color, mechanical values are recorded over aging time. The lifetime of polymers depends on the chemical structure, testing temperature, type and concentration of antioxidants and other additives.

Photooxidation is tested by UV exposure, artificial (e.g. ISO 4892) or natural weathering (e.g. ISO 4582, ISO 4607). Change of the properties, e.g. visual appearance such as chalking, crazes, loss of gloss, and mechanical properties such as tensile strength, elongation, impact strength are measured. The light sources of artifical weathering equipment match the terrestrial sunlight, a considerable acceleration is achieved versus natural weathering. Typical exposure sites for natural weathering are Florida, Arizona and southern France. For example 2760 hours of artificial weathering correspond to approximately 1 year Florida exposure when 50% retained tensile strength of PP tapes are considered as test criterion and when the PP is stabilized with 0.10% hindered amine stabilizer.

As a consequence of predamage from first processing and use, most plastic recyclates will not fulfill the requirements of the same or similar application if they are not upgraded by selected additives as outlined later. However, to understand the influence of the predegradation on plastic properties a few published examples should illustrate the effect on the most important polymers in a simplifed way with focus on stability and performance:

- Polypropylene (PP): The molecular weight of PP (and PP copolymers) is significantly reduced by processing and aging with a direct impact on the mechanical properties. Long-term applications of PP recyclate from, for example, battery cases result in inferior stability. 16 days until embrittlement at 135 °C were found for the recyclate versus 38 days for the virgin alternative. Virgin PP elongation at break was at 680%, whereas after five extrusions only 20% were recorded.

- Polyethylene (PE): Depending on the type (LDPE (low density), LLDPE (linear low density), HDPE (high density)) and the manufacturing catalyst PE tends mainly to cross-linking through aging. Most critical during outdoor exposure is crack formation and catastrophic failure. For example waste bins made from HDPE achieved a tensile impact strength of 404 kJ/m² whereas the recyclate attained only 291 kJ/m², a significant drop in mechanical properties due not only to aging but as well to a product mix from different manufacturers, use time and exposure. Furthermore, it was shown in experiments with LDPE and HDPE that alternating processing and aging caused considerably more severe degradation then either processing or aging alone.

- Polystyrene (PS): The main degradation path of PS recyclates is some molecular weight reduction and discoloration after thermal exposure, impact polystyrene suffers from cross-linking of the elastomer part.

- Polyvinylchloride (PVC): Predamage of PVC recyclate from rigid PVC applications such as window profiles or pipes is limited to the most outer part of the application, therefore the values of the mechanical properties after regrinding and reprocessing are often still acceptable. Less stabilized materials such as PVC bottles need addition of heat stabilizers, e.g. on the basis of Ca/Zn.

- Polyethylene terephthalate (PET): The PET recyclate, mainly from used bottles, shows a loss of molecular weight (virgin material: intrinsic viscosity 0.73–0.80 dl/g, recyclate less than 0.70 dl/g), caused by degradation through processing and hydrolytic degradation.

- Mixed plastics: Mixed plastics contain different types of plastics with different processing behavior and stability. Usually these plastics are not compatible (or thermodynamically miscible) with each other and the resulting properties are very often inferior to the properties of the parent polymers. The influence of plastic mixtures on the stability can be demonstrated in a model experiment: standard virgin HDPE, e.g. used in bottles achieved easily more than 100 days until embrittlement at 120 °C, the recyclate from 100% HDPE achieved still 73 days, taking a mixture of 5% PP and 95% HDPE reduced the stabiliy to only 18 days with further reduction at increasing PP concentration.

- Composites: Glass fiber reinforced polybutyleneterephthalate (PBT) composites showed in model experiments reduced impact strength (31.1 MPa vs. 37.6 MPa) and tensile strength (103 MPa vs. 134 MPa), resulting probably from a reduction in molecular weight. A similar result was found for glass fiber reinforced PPS. Other researchers confirmed that glass fiber reinforced PBT showed reduced tensile strength and elongation; however, improved values of Izod notched impact strength of the recyclate: the latter was attributed to inhomogeneities within the composite and fiber bundles which increase fracture energy. As the properties of composites are determined by the fiber type and dimension and by the interaction with the matrix it is likely that any composite will show inferior impact properties as recyclate independent of the structure if the fiber dimensions change during the recycling process.

COMMON TYPES OF RECYCLING

Almost all materials can be recycled. Recycling helps to reduce the number of materials that are thrown in the trash. These materials are processed and reused for various projects. Most importantly, recycling prevents the accumulation of waste on landfills, maintaining the natural beauty of the environment.

Today, public places, restaurants and companies are making it easier to recycle by placing labelled recycling containers in visible areas enabling everyone to recycle instead of throwing items in the trash.

Recycling offers many benefits for both the planet and people. It protects the earth, positively impacts the economy, conserves natural resources, reduces extraction of raw materials, and reduces energy use.

Paper

Tons of papers are used nationwide for printing and writing which are then thrown away as scrap. If the paper is not recycled, they contribute to greenhouse gas emissions and substantial garbage dumps at landfills. Paper is made from the pulp of trees.

When we recycle paper, we save a million trees on our planet and prevent deforestation. Today companies are using recycling paper for many of their printing needs such as business cards, general paper printing, writing proposals and creating memos. Even magazines, newspapers and books can be recycled.

Cardboard

Recycling just one ton of cardboard saves roughly nine cubic yards of landfill space. Cardboard is commonly used in households and companies. They are used for packaging and shipping of products. Both large and small cardboard boxes can be recycled even though they are easily biodegradable.

Many residential properties and companies offer a separate recycling container for cardboard boxes. However, there is an essential step before recycling cardboard boxes. They must be folded so that there is more space in the recycling container. This step also saves a lot of hassle for recycling personnel.

Steel

Steel is known to be the most recycled material on the planet because it is 100% recyclable. The advantage is that steel does not deteriorate during the recycling process. By recycling steel, we save almost 74% of the energy needed to make steel from raw materials. It also reduces air emissions, water pollution, and the extraction as well as the processing of iron ore.

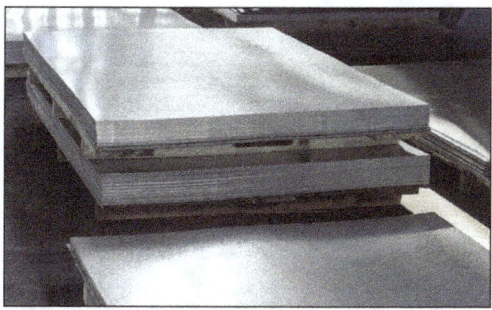

Aluminum Cans

Aluminum is used for beverages such as soda and fruit cans. There are billion aluminum cans manufactured and used every year. The only problem with manufacturing aluminum cans is that it demands large amounts of energy to mine raw materials. Producing aluminum cans from recycled cans only take about 5% of the energy resulting in 95% of energy savings. Apart from energy savings, there are cost savings as well. The cost of recycling aluminum is very low compared to the production of new aluminum.

Glass

Similar to steel, glass is 100% recyclable. When recycled, glass does not lose its quality or purity which means they can be recycled over and over again. The recycled glass prevents emissions, saves energy and reduces consumption of raw materials. Glass takes millions of years to break down at the landfill whereas, it takes a few days for recycled glass to be made into new glass containers or bottles.

Wood

Wood is wasted if it ends up at the landfills although biodegradable. However, if recycled, wood can be reused as building materials or for paper production. Wood is one of the most valuable recyclable material as it can be transformed into a wide range of secondary products. Recycling wood resources reduce the need to chop trees. The wood salvaged is generally free from contaminants. The wood is crushed into small chips used to make new products.

Plastic

Recycling plastic should be taken seriously as today it is one of the most significant solid waste occupying most landfills and oceans. Plastic takes between 500 to 1000 years to decompose clogging valuable landfill space and contaminating our oceans. With the recent increase in consumption, plastics have been seen floating on the surface of the ocean harming the ecosystem. By recycling plastics, 80% of the total energy that goes into manufacturing new plastic products is conserved, leaving our oceans and landfills free from solid waste.

Fabric and Textiles

Recycling textiles and fabric is easy. Old clothes or other types of fabrics can be reused to make new garments, linens, draperies and even cleaning materials. There are lots of usage for these materials. Textiles can take hundreds of years to decompose which is why they are better off being recycled and reused immediately. Lately, many countries are setting up clothes collection bins across different cities.

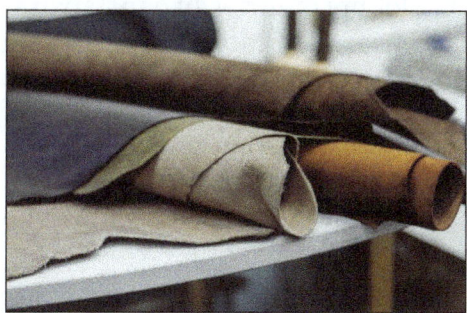

These types of recycling bins are placed in strategic public areas such as parking lots or supermarkets where people can drop off their old clothes on their way to do their grocery shopping. People buy and change clothes often which is why it is essential to try to reuse the fabric, instead of the fabric occupying massive amounts of landfill space.

ECONOMIC ASPECTS OF RECYCLING

Over the last years, environmental issues have gained remarkable significance as a result of the high level of environmental awareness in society and also due to the widening of the scope of the environmental legislation and regulation. One of the major concerns within this context is the high rate of packaging waste generation. Nowadays there is a variety of technologies to deal with waste treatment among which recycling is one of the most preferable. Recycling allows waste to be converted in resources and thus its economic value is prevented to be buried in landfill. Back in 2000, the World Business Council for Sustainable Development (WBCSD) proposed improving recyclability as a critical aspect for a sustainable development in business since this end-of-life treatment was foreseen as one of the most sustainable. Furthermore, international legislation regarding end-of-life of waste is pushing stakeholders to adopt more sustainable strategies in the long run. For instance, the European Union Directive 2008/98/EC establishes the five-step waste hierarchy, in which the most sustainable step is the waste avoidance, followed by reusing and recycling. Up until now, waste is taken into account as a product since it is possible to obtain economic benefit from it. After those treatments, recovery (including waste-to-energy) and landfill are considered as the last steps. The day when waste can no longer be avoided and we reach the maximum reuse rate possible, recycling will be the most preferable strategy. This approach is influenced not only by the point of view of the potential economic value of waste, but also by its eco-efficiency.

The benefits of recycling could be summed up in energy savings, natural resource conservation and reduction of waste disposal to landfill. Energy savings of recycling have been demonstrated in a wide range of materials. The fact of avoiding the use of virgin materials allows reducing natural resources consumption and thus preventing problems to future generations. Moreover, the high quantities and extensions of landfills is a serious disadvantage in a lot of countries like in Japan where land is scarce and it is hard to locate new final disposal sites. Moreover, a great amount of landfill sites are currently full and other solutions should be searched.

Thus, the recycling advantages from an environmental and an economical point of view are influenced by different elements. Environmental and economic life cycle analysis in combination with market studies can be used for the definition of the most suitable waste treatment process in a specific situation.

Markets for Recycled Materials

The market is one of the main aspects to be taken into account to decide whether a recovery process is economically viable. Indeed, there is a vast variety of influential indicators in the market and the prices of the material tend to fluctuate strongly over time.

Both market situation and prices ought to play a decisive role when it comes to choosing the appropriate waste treatment process in a certain context. This is due to the fact that the recycling process depends greatly on the efficiency and effectiveness of the previous and posterior processes needed. Example of such processes are collection of waste, transport, separation, conditioning, as well as post-treatments which are necessary for the transformation of the obtained product to a marketable product.

Obviously, another aspect that should be taken into consideration is the recycled material market price. For instance, as example for plastic waste, the selling price of the recycled material depends greatly on the price of virgin polymer, which is linked to the crude oil price as well as the electricity cost. On the other hand, the prices of the plastic waste bought by recyclers are rarely influenced by the oil price and financial profit influences it more than environmental concern (EuPR, 2010). The price of industrial and consumer waste polymers appears to be modified by:

- The export of materials, above all to emerging countries.
- High demand for high-quality recovered plastics.
- Collection and sorting companies trying to boost their income.
- The plastic characteristics regarding cleanliness, colour, etc.

It should be stated that sometimes the value of waste materials involve negative figures. In other words, the user is paid for acquiring the recovered materials. Steel mills accept waste plastic against payment and use them, after pre-treatments, as a reduction agent in a mixture of pulverized coal in blast furnaces. Thus, they avoid buying more expensive materials as raw material in mills.

The law of supply and demand is an influential factor on the material price setting. For instance, material exportations to other areas of the world like Asia influence the price. When the demand in the Asiatic market is high, the availability of such material in Europe decreases and consequently prices tend to increase. On the contrary, high availability of either virgin or recycled material usually causes a reduction of prices.

The geographical location plays also an important role in the price setting caused by differences in prices of machinery, materials, and production and labour costs. For example, production and labour costs in emerging countries are known to be lower than in Western countries, but inaccessibility to raw materials could revert to a less advantageous position in the global market. According to statistics, developed countries produce higher waste quantities per person, although high-populated emerging countries like China could vary this trend. Anyhow, recycling constitutes not only a challenge in environmental terms, but also a great opportunity to make profit from it.

Mixtures of recycled and virgin materials can be used instead of purely virgin material in manufacturing processes. Indeed, there are numerous studies showing that the addition of certain amounts of recycled material to virgin, does not damage the properties and characteristics of the resulting material.

However, there are cases where recycling might not be the best way to deal with waste. For example, when the price of the recycled material is highly priced with regard to the virgin material, it would be more profitable to choose the recycling option. On the contrary, in periods where the recycled material is low-priced would rather be more beneficial to focus on waste-to-energy technologies and obtain profit from the sale of energy.

On the other hand, recycling could have significant economic impacts as it replaces materials commonly obtained, transported and manufactured outside a specific region with materials collected and processed usually within the region.

Therefore, the selection of the recovery process that performs best in a given context requires a multi-perspective approach including issues such as financial cost, environment, market, supply, demand, etc.

Paper and Board

The major paper producers in the world are Asia, the CEPI countries 1 and North America. These figures are closely related to population and level of industrialisation of countries. The paper production decreased as a result of the world economic crisis in 2008 and still continued in 2009. This fact was partially balanced by the pushing of the Asiatic countries, especially China. Nevertheless, pressure on the recovered paper sector set up by governments appears to explain part of the statistics and the use of this material in the paper and board industries at the expense of virgin fibres. In 2010 the negative tendency was reversed and the utilisation of recovered paper was increased. On the other hand, 2009 was the fourth year in which generation and consumption of recovered paper outstripped virgin fibres ones, but it should not be forgotten that virgin fibres are indispensable in producing high quality paper. Furthermore, the use of fresh fibres is required to strengthen the mixture of recovered paper during the papermaking process. This is due to the degradation that fibres suffer in the process of recycling and the use of recycled fibres usually varies between three and eight times depending on the quality of the recycled paper.

Taking a look at the figures for the consumption recovered paper 2 in Table, the influence of Asia is stated. The differences between collection volumes and apparent consumption are explained by quantities in stock by collectors and consumers at the end of the year and a certain quantity of paper cannot be taken advantage like the toilet paper.

Table: Consumption of recovered paper in 2009.

	Collection (tonnes)	Imports (tonnes)	Exports (tonnes)	Apparent consumption (tonnes)
Asia	83,108,000	35,980,000	7,710,000	111,334,000
Europe	62,980,000	12,950,000	24,650,000	49,250,000
North America	49,900,000	1,650,000	20,880,000	30,670,000
Latin America	10,020,000	2,105,000	670,000	11,455,000
Australasia	3,340,000	160,000	1,470,000	2,096,000
Africa	2,140,000	90,000	60,000	2,172,000
Total	211,488,000	52,935,000	55,440,000	206,977,000

By countries, the Table shows the importance of Asia and particularly China as it comes to imports of recovered paper.

Table: Major importers for recovered paper.

Imports of recovered paper		
Country	2008 (tonnes)	2009 (tonnes)
China	24,200,000	27,500,000
Germany	3,556,000	2,860,000
Netherlands	2,472,000	2,964,000
Indonesia	2,080,000	2,290,000
India	1,755,000	2,135,000
Belgium	1,500,000	1,536,000
Mexico	1,436,000	1,510,000
South Korea	1,307,000	1,120,000
Austria	1,305,000	1,190,000
Thailand	1,218,000	970,000
Spain	1,170,000	902,000

Although, most of the recovered paper is consumed in the country of origin, the exports are measured in millions of tonnes and represent a huge international market since is the main material used to make new paper such as newsprint for newspapers and cardboard for packaging. Asia remains the most significant importer area due to the constant development of its paper and board industries. Regarding China, a fact that normally is not considered is that over the last years the Chinese government has encouraged to close many traditional mills because of uneconomic or focus of serious water pollution problems. In return, modern factories with highest capacity have been built principally in the east coast. As a result, the China´s capacity for processing recovered paper has been increased and estimations assure that this fact will continue in the next years. Thus Chinese producers arbitrate the world market export pricing.

As far as recovered papers prices are concerned, the price of selling them is rather dependant of their quality and desirable uses. For instance, according to EN 643 "European List of Standard Grades of Recovered Paper and Board", recovered paper grades can be divided as mixed grades (1.01), corrugated and kraft (1.04), magazines (1.06) and deinking paper (1.11). As a summary, in the figure the price of recovered newspapers is shown. One can state the extremely variations from both perspectives, the modifications in a country and the comparisons between countries.

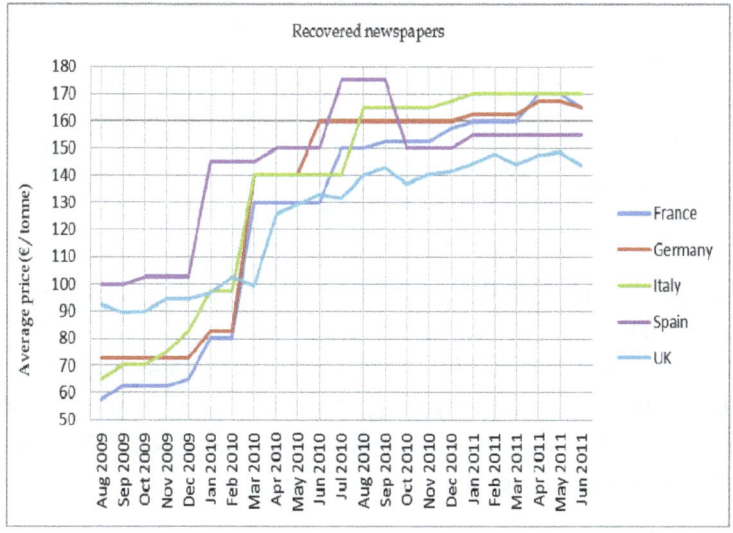

Average price of recovered newspaper for selected countries.

Plastics

Plastic waste can be recycled by mechanical and chemical recycling. Not all plastics can be recycled since there are chemical difficulties that make it impossible or with undesirable results. Mixed plastic packaging is not usually recycled due to problems in the separation stage and it is under research currently as it comes to collection and processing. Nevertheless, there are initiatives in order to adapt it to the industrial scale. Thus, the plastics which can be recycled are PET (polyethylene terephthalate), HDPE (high density polyethylene), LDPE (low density polyethylene), PP (polypropylene), PVC (polyvinyl chloride), PS (polystyrene) and others. The new uses range from jackets, coat and food packaging for PET. HDPE can be recycled in order to make tables, roadside curbs, benches, lorries cargo liners, trash receptacles, etc. LDPE is used as shop bags after the consumer stage. Recycled PP is suitable as small beans and battery boxes. Recycled PVC is appropriated for making weep pipes. PS can be converted in flowerpots. Furthermore, uncommon products from recycled plastics are being studied in order to add value to these materials. Examples of such innovations constitute the intelligent textiles.

World plastic production had been increased from 1950 to 2008. In 2008 the economic crisis made total tonnes of produced plastic to drop, but it seems like it is slowly recovering in 2010. The higher demanders of plastic are the NAFTA countries 3, Western Europe and Japan. Focussing on the European market, the demand for plastic was in 2009 as shown in Table.

Table: Plastics demands by converters in 2009. Breakdown by types.

Polymer	Demand in percentage (%)
LDPE	17
HDPE	12
PP	19
PVC	11
PS	8
PET	8
PUR (Polyurethane)	7
Others	18

Plastic recycling has been increasing its tonnage by around 11% per year over the last 10 years in Europe. Nevertheless, as stated above, in 2009 this growth fell to 3.1% as a direct impact of the economic crisis. This number is also explained by stronger activities of some packaging collecting and recycling systems and the increment of the exports outside of Europe, i.e. to the Far East. Derived from these facts, the quantity of plastic that end its life in landfill is reducing.

As it comes to the Asian market, the same tendency is observed. From 1990 to 2008 waste plastic trade was increased by 100 times as a result of the increasing demand of such material. High prices in oil were related to high prices in virgin plastics which caused an increment in recovered plastic demand. Thus, 80% of waste plastic was sent to Asia in 2007.

The most common plastics which are currently recycled are PET and PE as it comes to prices of recycled plastics. The data figures are usually presented varies with the type of plastic. As an example, the PET prices differ from whether the material is in form of bottles or flakes, if it is colourless or mixed colour, or if is suitable for food contact (food grade) or not. In figure, the prices for recycled plastics are presented for the British market.

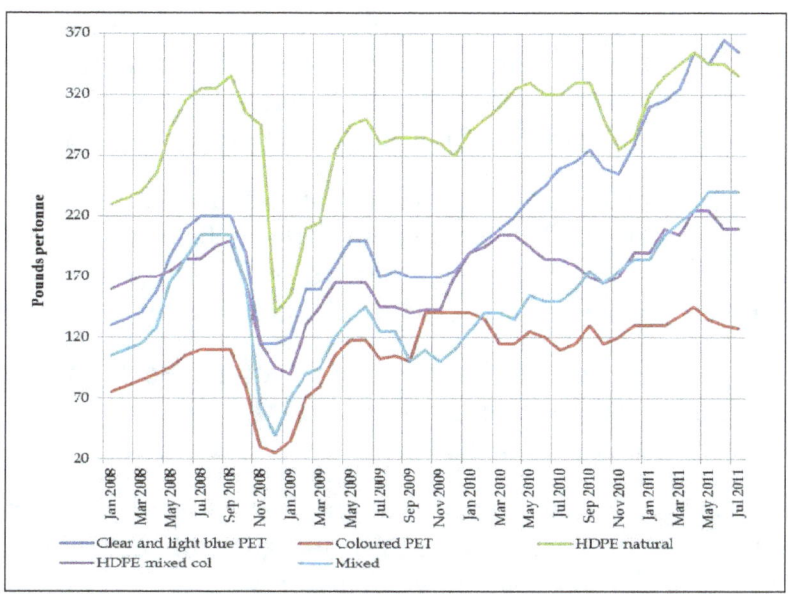

Average prices for different kinds of plastic bottles in the British market.

A clear drop in 2008 and the subsequent increment are shown for high quality sorted materials such as PET and HDPE bottles. High demand from Chinese processors, depreciation of currencies against the US dollar, increasing demand for food grade recovered plastics, tight domestic supply, related to the lower consumption due to crisis and higher virgin plastics prices can explain this behaviour. Plastic demand is strongly sensitive to consumer spending habits, and therefore, economic downturns. Nevertheless, a recovery is observed for every type of plastic. The line for mixed plastic bottles remains more horizontally than the other kind of recycled plastics because of uncertainties in their process of recycling and the need to study deeply their recyclability. On the other hand, and focussed on the plastic container sector, disruptions are influenced by mergers and acquisitions of companies especially in the USA due to large players that are fighting in a mature market.

Metals

The advantages of recycled metal versus virgin one, from an environmental point of view, are the lower use of natural resources and the savings in the energy required for its processing. Recycling saves 95% of energy in aluminium production, 85% in copper production, 74% in steel production, 60% in zinc production and 65% in lead production. In economic terms, this is translated to savings in the energy bill and the possibility of taking profits from an interested market. Such importance lays on the metal market has raised their demand in China, India and the Far East due to the development of the commercial, residential and industrial construction. Moreover, industrial machinery, cars and armament demand these materials. Metal scraps help the primary metals industry to achieve these needs not only in the developing countries, but also on the Western countries.

Steel is by far the most-recycled material in the world, followed by paper and aluminium. As it comes to economic figures, the main exporters in 2010 were USA (20.56 million tonnes), the European Union with 18.97 and Japan (6.47), whereas the main importers were Turkey (19.19 million tonnes), South Korea with 8.09 and China (5.85).

The steel scrap consumption in the European Union increased in 2010, whereas USA slightly reduced it in that year. The main reason for that behaviour is that electric arc furnaces (EAFs) only consume steel scrap because of saving in energy consumption, but were operating at lower rates. Moreover, the EAFs in USA utilized around 40% iron alternatives.

Asia had become a net centre of recyclable import with the 80% of the world's copper scrap imports, the 50% of the aluminium scrap and the 30% of iron and steel scrap in 2007. China has converted in the world leader in steel production.

Figure shows the prices paid by medium to large size recycling companies which receive material from local collectors. As shown in figure, the economic crisis has not influence the metal prices so deeply in comparison with the previous materials and prices are recovering slowly. This is due to the constant demand of these materials. For instance, in the aftermath of the recession the steel production in the world reached 1.412 billion tonnes in 2010, which represented an increase of 14.8% over 2009 and a new record. Only copper reduced its price in 2008 and roller coaster movements have been done in the last months. This uncertainty in the price is partially due to the crisis of the Europe sovereign debt, the downturn of the America´s economy and the process of restructuring the Chinese economy.

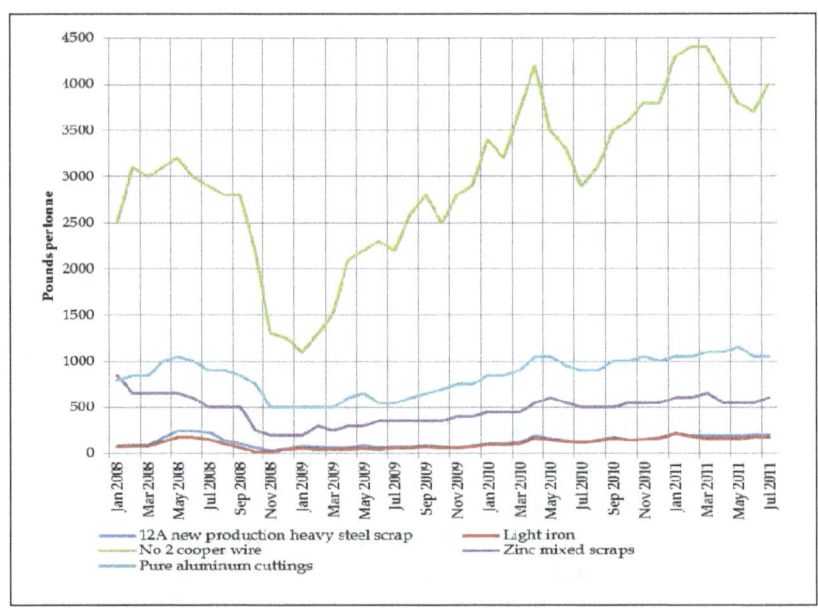

Average values of scrap metal prices in the British market.

Not surprisingly, the price of non-ferrous metals is appreciably incremented in comparison to the one of the ferrous metals. The main reason is the limited availability of those materials, so the unrestricted flow of non-ferrous scrap from country to country is crucial according to the law of supply and demand.

The precious metals constitute a very special case. The precious metals´ price is increasing almost continuously. They act as a safety refuge when global markets are unstable. Furthermore, when the price of oil is moving up and down, investors look for these kinds of metals with relative stabile value on the market.

Glass

The process of glass recycling is quiet interesting from an industrial and environmental point of view. Glass can be recycled endlessly without losses in quality or purity (GPI, 2010). Energy costs drop around 2.5% for every 10% cullet used in the manufacturing process. In addition, cullet makes the manufacturing mix less corrosive and reduces the melting temperature extending the furnace life. Moreover, glass recycling is a closed-loop system which does not produce additional waste or by-products.

Demand for cullet has grown over the last years with new glass processing plants and alternative markets under developing. Thus, the domestic market for recycled glass remains usually at high level and this fact makes the price paid for this material to be quite stable. In most countries this circumstance is added to the non-dependence of the glass industry on the export market. As a result, long-term contracts between local authorities, which are the main providers of glass for recycling, and glass recycling companies come with a good degree of security.

On the other hand, in the last years companies have adopted strategies in order to improve their environmental credentials as a way of making their green marketing better. Derived from this fact, actions like reducing the weight of their glass bottles in important percentages have been doing. 99% of glass containers in the world (in weight) belong to food packaging and it is easy to imagine the global avoided impact of these actions.

Imports and exports in the glass market are rare because of its high weight and the high fuel consumption associated to their transport which increases prices. Moreover, multinational companies have plants in several countries and the raw material recovered glass is mainly provided by the national market.

In the figure the prices of glass containers in the British market are shown. Completely mixed glass has the lower value due to strange colours obtained after melting. This fact makes it difficult to sell it since customers demand pure colours. Instead, it should be derived to alternative uses. Green glass has the third lower price because it is the most abundant in the United Kingdom waste stream. The clear glass is the most expensive since it is the most demanded one in this market. Nevertheless, internal issues could affect the glass price. For instance, in the figure the slump in prices from September to December of 2010 was related to the continual decline in Packaging Recovery Notes revenues.

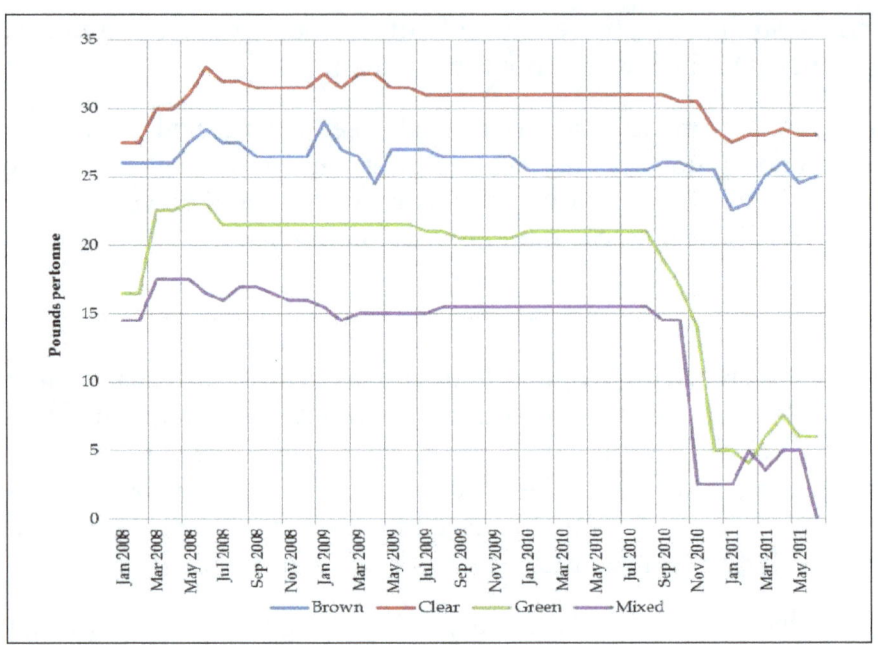

Average values of glass prices of containers delivered to collectors in the British market.

Apart from glass containers, alternative uses of recycled glass are coming out. Such new approaches include glass as grit blasting, use in road surfaces or water filtration.

Wood

As it comes to recycling of wood, two types of materials should be distinguished. i) High grade wood comes from clean white softwood, wooden pallets, timber offcuts, packaging crates and joinery waste. They should be free of paint and coverings. ii) Low grade wood includes plywood, doors and window frames, roof timbers, panel products and so on. Both kinds of materials come from construction and demolition, commercial, industrial and household sources.

The main market of recycled woodchip has been the panel board industry due to historical reasons. This material is used in the production of chipboard. Nevertheless, the wood industry has been searching for value-added markets and new uses have been discovered, i.e. animal bedding, equine surfacing, garden mulches markets have steadily developing. This fact has made wood industry to increase in the last years. The demand of woodchip is expected to continue growing since governments are promoting the generation of renewable energy.

In the figure, the price evolution of recovered mixed wood delivered to a wood recycler can be seen. Negative prices indicate that the recycler is paid for these materials. Nevertheless, this fact can change because of increasing demand of clean wood pallets and sawmill round wood towards small payments by suppliers. In addition, wood waste prices will vary considerably depending on cleanliness of the material, volume and location.

The Recycling Supply Chain

Recovered waste materials, before being recycled must be collected, transported and separated among other processes and as a result, the general performance of the recycling process depends greatly upon the efficiency and effectiveness of those "minor" processes. Similarly, from the economic perspective, the recycling process must be preceded from also efficient and economic sub-processes. Otherwise, high cost of recycling could overweigh the environmental benefits of recycling. In addition to the recycling process itself, a broader approach including the prior and posterior stages of recycling is needed.

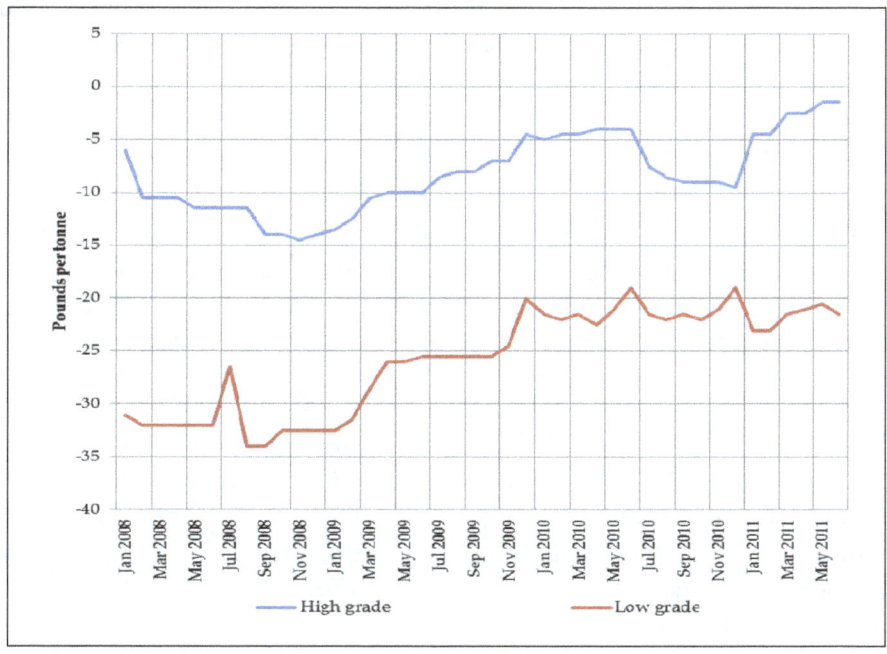

Average values of wood prices of mixed wood delivered to a wood recycler in the British market.

Collection

The municipal solid waste generated in the European Union was 514 kg per inhabitant and 719 kg in the USA in 2009. Such high quantities should be treated under the most convenient processes. The end of life of materials starts wherever is produced, whether it is generated in a house, store or industry. It is necessary to separate out waste in appropriate fractions according to the facilities in which they are going to be treated in order to make further stages more effective. Throughout this action waste is treated easily and economic costs are reduced, allowing making recycling a competitive treatment. Ideally, waste should be set apart as paper and board, plastic, metals, glass, wood, batteries, textile and organic. This can be done at source, having selective collection in the cities or at specific sorting plants, where all recyclable materials are sorted out and sold to recyclers. Nevertheless, the availability of these facilities varies from country to country and is close related to environmental policies.

Transport plays an important role in the collection stage. A wide range of vehicles can be used, but lorries have become as the most used in this issue due to their load capacity and the particularities of towns. Furthermore, these vehicles can sometimes incorporate a pre-waste disposal unit which enables compacting waste and making collection more effectively.

Sorting

Once waste is collected, sorting is a mandatory activity in order to have separated the different recyclable materials and to avoid undesirable materials which can decrease recycling efficiency and recycled material quality. For example, ceramic is considered a disadvantageous material when glass is recycled. Furthermore, separating a material in specific fractions (for instance different colours, etc.) allows taking advantage of differences in sell prices as has previously been shown in the elements studied. Paper and board´s costs differ on the quality of fibres and, at the same time, on the origin. Plastic should be separated according to the type of resin because some recycling processes are specific of each material. For instance, methanolysis only works with PET polymer. Glass is required to be sorting according to its colour. Coloured glass is unable to produce clear one and mixed glass presents the lower price. Metal values depend on their kind as we have seen and sorting is a must in order to separate them according to their nature.

The cost of sorting depends on the technique selected. For example, in case of using optical separators the purchasing cost of the equipment could be more expensive than carrying out a density separator, i.e. a pool with water and salt (to increase the density) or alcohol (to reduce it) is enough for separating plastics. However, this last treatment could lead to an increment in the final cost by considering also the price of energy in order to dry wastes as well as the substances used for adapting the water density and the water treatment after the process.

Then we have two flows: the materials which will continue with the recycling process and the sorted materials that can be sold in order to obtain revenues.

Shredding

Shredding is required before the recycling process and is considered a key pre-treatment since it facilitates further processes. Its main advantage lies on the volume reduction, which minimize the storage space as well as reduce transportation. Moreover, as it comes to glass, it allows reaching the melting point by using less energy and time, and improving furnace feeding. Small pieces increment their surface exposed and the energetic yield. Thus, less fuel is required in order to melt glass and metals. It is desirable that metals arrive at the processing facility in bales or briquettes since this way the efficiency of the remelting process is increased. Nevertheless, the most common is not to receive such material in this form and a shredder constitutes usually a stage in the recycling facility. In addition, it helps to liberate trapped contamination, upgrading the quality of materials like glass, metals or plastics and its value. On the other hand, the shredding of paper and wood is not high recommended since fibres will be damaged. Wood recycling requires shredding when the final use is to make agglomerate boards or smaller pieces.

Recycling Process

The recycling process differs depending on the material to be treated. The recovered paper is put in a pulper with water and is pulped with mechanical and hydraulic agitation in order to disintegrate paper into fibres. For purposes of deinking some chemicals are sometimes used, like deinking agents and NaOH. Contaminants are removed during the operation due to differences in physical properties. The pulp slurry is pumped from the pulper to hydrocyclones. The organic rejects are often burned in order to take advantage of their calorific value. In general, screening at lower consistency in order to separate undesired particles is more effective, but it requires additional machinery installations and its energy consumption of the process is increased. After that, a fractionator is used in order to separate the pulp in two fractions, creating a short-fibre stream and a long-fibre one, to apply different treatments. In the case of long-fibres, dispersion can occur with the finality of achieving better fibre-to-fibre bonding, strength characteristics, and to reduce dirty specks in size. Furthermore, refiners improve optical and strength characteristics, but its main disadvantage is the impressive energy consumption. Continuing with the industrial paper recycling process, the mixture enters in the paper machine after a cleaning substage and a fine screening. Moreover, a flotation deinking stage is recommended when recycling paper so that ink is removed and a better brightness is followed. In addition, bleaching chemicals can be added before entering at storage tower. Finally high quantity of waste water is produced during the process and should be conveniently treated in order to reduce pollution.

In case of plastic recycling, after sorting and shredding, a washing step should take place in which impurities are removed. Next, the melting process is done throughout an extruder which applies heat by friction. It also allows homogenizing and filtering in order to produce high quality recycled material. In the end, the pellet conformation is produced by a pelletizer.

The recycling process of metals consists of a melting process. The temperatures should reach from hundreds to thousands degrees, but alloys can reduce such high values, high energetic costs, to a more reasonable one. There are particularities in the melting process due to the wide range of recyclable metals and stages for overcoming potential problems. For example, as it comes to aluminium alloys, a degassing step is necessary to reduce the amount of hydrogen in the liquid metal. High hydrogen concentration could result in gas porosity that deteriorates mechanical properties. Continuing with the process, the molten metal is poured into molds. Once the metal is solidified, it is removed from its mold. A degating stage is necessary in order to remove head, runners, gates and risers from the casting. For this purpose, cutting torches, bandsaws or ceramic cutoff blades are used. This metal must be remelted as salvage increasing, thus, the yield and reducing the costs. Surface cleaning is needed for a better presentation. Usually, sand or other molding media might be adhered to the casting and metal is cleaned using a blasting process. In other words, a granular media is propelled against the surface of the casting. The media, propelled by compressed air for instance, strikes the surface at high velocity and tears any impurity of the surface. Finally, grinding, sanding or machining steps are done in order to achieve the desired dimensional accuracies, physical shape and surface finish, as well as painting in case prevention of corrosion was needed and improve visual appeal.

Once again, melting is also the main step in the glass recycling process. Cullets are melted in huge furnaces. Decolorizing and dyeing is the next stage. Firstly, oxidizing of the melted glass cullet is required. For green glass, the colour turns from green to yellow-green and manganese oxide is then mixed until a grey colour appears. For brown glass, zinc oxide is added to oxidize it to blue or green cullet. If clear recycled glass is required, erbium oxide and manganese oxide are added to help clear all the colours from the glass cullet. In the end, the recycled glass is moulded into the final product.

Taking into account that contaminants have been removed before, the recycling process for wood consists in a two-steps process in which wood is introduced in a tub grinder, horizontal grinder or wood chipper. Wood is grinded into chips which are ready to be sold for use in particle board, chipboard, pulp and paper products, animal bedding, mulch, biomass fuel and compost. In the panel board industry, an adhesive is necessary to stick wood pieces together.

Distribution of Recycled Material

After recycling, two additional steps should be also considered: packaging and transport to customer. As it comes to packaging, the type is influenced by the transported material. Paper is carried in coils, and for plastic pellets sacks are usually required. Moreover, metal and cardboard can be transported with strapped sheets. Glass recycling and manufacturing are done in the same facilities and transportation to glass maker is not required, although transport from glass maker to filler is needed. Here, packaging entails of palletizing and shrink wrapping. Initially, the distribution stage does not modify substantially the purchasing cost of the recycled material, but it is greatly influenced by the distance, weight and way of transport.

Economic Evaluation of Recycling

The aim of the economic evaluation in recycling processes is, on the one hand, to assess the economic impact of recycling and on the other hand, to identify weak points, or the less economically efficient stages of the process, in order to be improved. This data is valuable for decision support from a public and private point of view. In regard to public organisms, it allows to select the best waste treatment alternative. As it comes to private companies, it also provides information about the recycling process´ profitability and recovery.

The economic evaluation of recycling processes can be determined through several methods. Some of the most relevant are the Life Cycle Costing (LCC), the Cost-Benefit Analysis (CBA) and Input-Output methods. LCC is a method which follows a life cycle perspective. All the inputs and outputs are determined in economic units and several types of LCCs can be found. CBA has been developed for major public investment plans and compares the total expected costs against the total expected benefits. This way the difference between costs and benefits is determined and measured. The Input-Output method is a linear model in which the interdependencies between links in a chain are presented on the basis of the output of one industry is the input of another. This quantitative economic technique is used to compare branches of national economy or between competing economies.

Despite the fact that the economic evaluation of recycling can be worked out throughout diverse methods, in this chapter LCC will be the most selected economic method since it is a precise and useful tool suitable for economic evaluation of recycling.

Life Cycle Costing

LCC analysis, which is a cost management method, is carried out on existing products and is used to monitoring and managing costs over the product or process life cycle. From a time perspective, it can be used for comparing past alternatives or future ones.

Environmental LCC analysis is a methodology that allows:

- To identify the processes that are most relevant for the overall cost.
- To compare life cycle costs of alternatives.
- To detect direct and indirect (hidden) cost drivers.
- To identify trade-offs in the life cycle of a product.
- To use the full costing to identify new products.

One of the first steps to carry out a LCC is to determine the functional unit (FU) to which all the costs will be referred. Examples of FU in a recycling process could be one tonne of recycled product. However, the FU should be adapted to the specific case considered in order to reflect the real costs. For instance this means considering the same time frame for all costs and including, if necessary the effect of inflation. The system boundaries should also be defined. System boundary limits which stages, inputs and outputs are taken into account in the LCC. Ideally, considering the whole life cycle is the best, but sometimes it is impossible due to missing data. In this case, one can make a bibliographic search to overshadow the handicap, although in case of not finding the data the system boundaries should be modified. All the data will be expressed in economic terms and converted into present-value costs by using discount rates. Another piece of advice highly recommended is to break LCC in cost elements so that it is easy to use. Not all costs should have a great impact on the final results, but breaking costs constitutes a good starting point aim at considering those that contribute most to the total cost.

There are many ways to classify the costs involved in a LCC. One of them is the approach of according to which the costs are:

- Internal costs (IC) along the life cycle of the products (i.e.: production, use or end-of-life expenses), which are the costs for which the company is responsible over a period of time. These type of costs include:

  ○ Conventional costs (CC): direct costs borne by the company when manufacturing a product (i.e.: raw materials, electricity, transport, etc.)

  ○ Hidden costs (HC): general costs related to license expenses, waste management costs, etc.

  ○ Less tangible costs (LTC) which are often not included in the company accounts due to their probabilistic nature. These costs include expenses on marketing, improving the image of the product, safety measures for workers, etc.

- External costs (EC) that are envisioned to include monetized effects of environmental and social impacts not directly billed to the company, consumer or government. These costs are also called "externalities" on life cycle management forums (like costs related to depletion of natural resources, impact on human health). Quantifying of negative effect of external costs is a critical issue in LCC practice. According to Kloepffer (2008) external costs to be expected in the decision-relevant near future (e.g. cost occurring in the future due to legal requirements in order to fight against climate change or special requirements for radioactive waste) are difficult or even impossible to estimate. Several attempts for the monetisation of external costs have been done, i.e. monetisation of emissions, road congestion and noise negative effects.

The definition of the internal costs as the costs "directly borne by an individual or organization in supplying or consuming a product". And the external costs were defined as "the market costs, not directly borne by an organization in terms of costs of labour, capital, and taxes, but as costs for purchases from other firms in the system, covering the internal costs of these other firms".

Consequently the LCC can be calculated as follows:

$$LCC = IC + EC = (\Sigma\ CC_i + \Sigma\ HC_i + \Sigma\ LTC_i) + \Sigma\ EC_i$$

Figure represents the stages of the life cycle of a product. As can be seen, the life cycle of a product can be a close-loop in case recycling is used as a waste management. Costs in LCC correspond to the stages in life cycle. This way, finding the most disadvantage phases is easy.

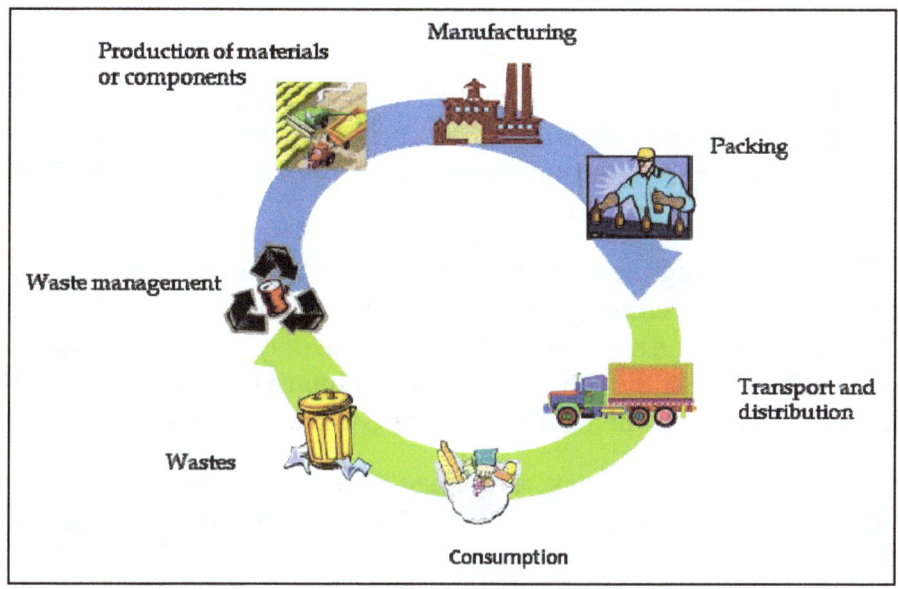

General life cycle of a product.

The SETAC-Europe Working Group on Life Cycle Costing has defined three types of LCC, depending on the costs considered in the study: conventional life cycle costing, environmental life cycle costing and societal life cycle costing.

Conventional LCC

The assessment includes only internal costs excluding on some occasions the End of Life (EoL). The perspective usually refers to one actor: the manufacturer, the user or the consumer. The conventional LCC is similar to the profit-loss account carried out by companies and does not include social impacts. It is usually not associated with separate LCA results.

Environmental LCC

Environmental LCC may be defined as "an assessment of all costs associated with the life cycle of a product that are directly covered by any one more of the actors in the product life cycle (supplier, producer, user/consumer, EOL-actor), with complimentary inclusion of externalities that are anticipated to be internalized in the decision-relevant future".

The assessment includes internal cost plus external costs expected to be internalized. Complete life cycle is taken into consideration and the perspective used may refer to one or more actors. Environmental LCC takes into account the financial costs associated to environmental impacts from LCA. The fact of using the same functional unit allows the comparison between LCA and LCC. This type of LCC allows the assessment of costs along the life cycle of a product maintaining the economic pillar of sustainability separately from the environmental and social pillars.

Societal LCC

The assessment includes both internal and all external costs and complete life cycle is considered. The perspective refers to society including governments and considerers both, the present and the long-term future possible situation. Societal LCC includes all of Environmental LCC and additional external costs.

The main difference among the three types of LCC is that only environmental LCC follows a functional unit as a reference unit for the economic assessment. This aspect is extremely important, since the environmental LCC is the only kind of LCC that follows the same approach like LCA. Although not existing an international specific methodology for carrying out LCC studies, the reference can be used to carry out this kind of analyses.

Once the types of LCC and costs are explained, a review of some case studies will be carried out in order to know more about this kind of analysis and better understand how to interpret the results.

SCRAP

A scrap is typically a small item that originally was part of something larger, like a scrap of fabric that was once part of a larger piece. Scrap can also describe something that's no longer useful.

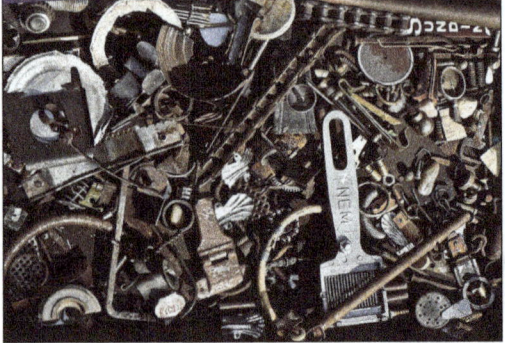

Piles of scrap metal collected for the World War II effort.

Collection of leftover scrap metal items.

Scrap consists of recyclable materials left over from product manufacturing and consumption, such as parts of vehicles, building supplies, and surplus materials. Unlike waste, scrap has monetary value, especially recovered metals, and non-metallic materials are also recovered for recycling.

Processing

The "organized chaos" of a scrapyard.

Scrap metal originates both in business and residential environments. Typically a "scrapper" will advertise their services to conveniently remove scrap metal for people who don't need it.

Scrap is often taken to a wrecking yard (also known as a scrapyard, junkyard, or breaker's yard), where it is processed for later melting into new products. A wrecking yard, depending on its location, may allow customers to browse their lot and purchase items before they are sent to the smelters, although many scrap yards that deal in large quantities of scrap usually do not, often selling entire units such as engines or machinery by weight with no regard to their functional status. Customers are typically required to supply all of their own tools and labor to extract parts, and some scrapyards may first require waiving liability for personal injury before entering. Many scrapyards also sell bulk metals (stainless steel, etc.) by weight, often at prices substantially below the retail purchasing costs of similar pieces.

A scrap metal shredder is often used to recycle items containing a variety of other materials in combination with steel. Examples are automobiles and white goods such as refrigerators, stoves, clothes washers, etc. These items are labor-intensive to manually sort things like plastic, copper, aluminum, and brass. By shredding into relatively small pieces, the steel can easily be separated out magnetically. The non-ferrous waste stream requires other techniques to sort.

In contrast to wrecking yards, scrapyards typically sell everything by weight, instead of by item. To the scrapyard, the primary value of the scrap is what the smelter will give them for it, rather than the value of whatever shape the metal may be in. An auto wrecker, on the other hand, would price exactly the same scrap based on what the item does, regardless of what it weighs. Typically, if a wrecker cannot sell something above the value of the metal in it, they would then take it to the scrapyard and sell it by weight. Equipment containing parts of various metals can often be purchased at a price below that of either of the metals, due to saving the scrapyard the labor of separating the metals before shipping them to be recycled.

Resources

Loading scrap gondolas in Eugene.

Scrap prices may vary markedly over time and in different locations. Prices are often negotiated among buyers and sellers directly or indirectly over the Internet. Prices displayed as the market prices are not the prices that recyclers will see at the scrap yards. Other prices are ranges or older and not updated frequently. Some scrap yards' websites have updated scrap prices.

In the US, scrap prices are reported in a handful of publications, including *American Metal Market*, based on confirmed sales as well as reference sites such as Scrap Metal Prices and Auctions. Non-US domiciled publications, such as *The Steel Index*, also report on the US scrap price, which has become increasingly important to global export markets. Scrap yards directories are also used by recyclers to find facilities in the US and Canada, allowing users to get in contact with yards.

With resources online for recyclers to look at for scrapping tips, like web sites, blogs, and search engines, scrapping is often referred to as a hands and labor-intensive job. Taking apart and separating metals is important to making more money on scrap, for tips like using a magnet to determine ferrous and non-ferrous materials, that can help recyclers make more money on their metal recycling. When a magnet sticks to the metal, it will be a ferrous material, like steel or iron. This is usually a less expensive item that is recycled but usually is recycled in larger quantities of thousands of pounds. Non-ferrous metals like copper, aluminum, and brass do not stick to a magnet. Some cheaper grades of stainless steel are magnetic, other grades are not. These items are higher priced commodities for metal recycling and are important to separate when recycling them. The prices of non-ferrous metals also tend to fluctuate more than ferrous metals so it is important for recyclers to pay attention to these sources and the overall markets.

Hazards

Great potential exists in the scrap metal industry for accidents in which a hazardous material present in scrap causes death, injury, or environmental damage. A classic example is radioactivity in scrap; the Goiânia accident and the Mayapuri radiological accident were incidents involving radioactive materials. Toxic materials such as asbestos, and toxic metals such as beryllium, cadmium, and mercury may pose dangers to personnel, as well as contaminating materials intended for metal smelters.

Many specialized tools used in scrapyards are hazardous, such as the alligator shear, which cuts metal using hydraulic force, compactors, and scrap metal shredders.

Scrap Metal

Scrap metal is the combination of waste metal, metallic material and any product that contains metal that is capable of being recycled from previous consumption or product manufacturing.

Scrap metal can originate from commercial and residential use. Whether it's ferrous or non-ferrous metal, the processing of this into vital secondary raw material for the smelting of brand new metals is absolutely key.

These scrap metals have a high market value, with their ability to be re-used again and again. For instance, electricians might have wires and electrical equipment, plumbers are likely to have used copper piping and brass fixtures and even construction firms will have beam upon beam of steel that could be quite literally given a new lease of life.

But all too regularly these are tossed into the dump due to lack of knowledge and sources for metal recycling.

Determining Non-ferrous and Ferrous Metals

Before recycling any metals, the first important step is to determine whether a metal is ferrous or non-ferrous. This is a very straightforward process and requires only a common magnet. If the magnet sticks to your metal, it is a ferrous metal. If the metal does not stick to your magnet it is a non-ferrous metal.

The most valuable scrap metals for recycling are non-ferrous; the most common of which are those that do not contain iron and are more resistant to corrosion, including copper, brass, aluminium, zinc, magnesium, tin, lead and nickel.

Ferrous metals are less valuable to metal recyclers but they will still recoup some value if you have enough of it. This includes metals such as steel and iron. Steel can be found in so many places; from cars to chairs, cabinets, shelving and more.

The Most Valuable Non-ferrous Metals for Recycling

- Brass - Easily found on door handles, light fittings, keys and plumbing fixtures, brass is one of the most common yet in-demand non-ferrous metals. Often yellow in colour with a hint of red, brass is a combination of zinc and copper that can be extremely dense, increasing its value in pure weight alone.

- Aluminium - Yet another metal that's often found in so many places around a regular home, aluminium can be recycled and re-used in an alternative guise within a matter of a months. Empty drinks and food cans are the most common places to find this metal, but areas such as guttering, siding, internal and external door and window frames are also good places to look. Aluminium is such an attractive metal for recyclers as the process saves 80 per cent of the energy that was used to make it in the first place.

- Copper - Another common metal regularly found in the structure of homes across the country, copper is also very valuable to recycle and very much in-demand at scrap yards. If the copper itself is in good condition it will be reddish in colour, but more worn copper fixtures and fittings will appear dark brown and sometimes green in places. It's a versatile metal which means it is regularly used as plumbing pipes, as a roofing material for guttering, with common electric wires and even inside air conditioning units.

SUSTAINABLE SOLID WASTE RECYCLING

Nowadays, overpopulation and rapid development of industries and lifestyle lead to an increase in the consumption of natural resources and reduction of their resource. On the other hand, humans have always produced waste and disposed it in some way, which influence the environment. Therefore, the increase in waste that was generated by the industrial factories and the human activities needs to be managed. For this reason, scientists have discovered new types of engineering that include sustainable engineering and green engineering to reduce energy and natural resources consumptions. The idea of sustainability has a quantifiable unit, which refers to three pillars of social, environmental, and economic. They focus on the environmental policies, which increasingly require the reduction, reuse, and recycling of waste for contributing to closing the loop of material use throughout economy by providing waste-derived materials as inputs for production.

Sustainable manufacturing process and solid waste management are used for conserving valuable natural resources, preventing the unnecessary emission of gas and protecting public health. The main goals include reducing environmental impacts and offering economic opportunities. Solid-state recycling process becomes an effective and powerful methodology to realize the green state forming from recyclable waste to useful parts. The developed process can be considered as a typical green forming or environmentally manufactured process. It has many benefits including simple, cost-effective, and energy saving and can be clean recycled as it does not harm the environment.

The application of plastic materials and their composites continues to grow rapidly due to their low cost and ease of manufacturing. Therefore, a high amount of waste plastic is being accumulated, which creates a big challenge for their disposal. Disposal the sustainability of plastic for a wide variety of application, organizations are faced with the growing problem of finding alternative methods for disposing a large volume of waste packages. Disposal of plastic waste in environment is considered to be a big problem due to its very low biodegradability and presence of large quantity. Moreover, different types and sizes of metal chips are produced during manufacturing process of metal products. The generated chips had been recycled by traditional methods that include remelting and casting processes, which lead to loss of parts of chips due to their oxidation because of their size and weight. The traditional recycling process becomes an expensive method because it consumed high energy and generated high pollution. Furthermore, energy conservation and environment preservation are a challenging task worldwide. Therefore, sustainable manufacturing process is a promising technology to reduce the waste and cost as well as reducing the usage of primary natural resource by developing and improving lightweight materials. Solid-state metal conversion is one of the most important processes that can be used to eliminate the required energy for melting due its ability to produce solid parts directly from solid chips. Sustainable development is the development that meets the needs of the present without compromising on the ability of future generations to meet their own needs.

Sustainability

Humans that require everything for their survival and well-being depend directly or indirectly on the natural environment. Our health, economy and security are required high quality of environment. Sustainability is being used by international organizations as a common approach to address the three sustainability pillars that include social, environmental, and economic issues. The potential economic value of sustainability is recognized to not merely decrease environmental risks but also to optimize the social and economic benefits of environmental protection. Sustainability is used to create and maintain conditions under which humans and nature can exist in productive harmony that permit fulfilling the social, economic and other requirements of present and future generations.

Sustainability has been applied in the field of engineering, manufacturing, design, technology, economics, environmental stewardship and health. Therefore, sustainable development becomes a key objective in human development due to increasing human activity. For that reason, sustainable manufacturing process requires balancing and integrating economic, environmental and societal objectives.

The growing identification of sustainability as both a process and a goal ensures long-term human well-being. The recognition that current approaches for decreasing existing risks, however successful are not capable of avoiding the complex problem. But current and future human generations at risk due to overpopulation, the gaps between rich and poor, reduction of natural resources, biodiversity loss and climate change will reduce. Human beings are at the center of concern for sustainable development. The principle makes clear that human well-being and quality of life is the objective of sustainability. Therefore, sustainable is defined as meeting the needs of the present without compromising on the ability of future generations to meet their own needs. It is improving the quality of human life while living within the carrying capacity of supporting ecosystems, through vague conveys: the idea of sustainability having quantifiable limits.

A large number of tools can be applied to address component parts of an analysis. Several principles are important in applying the suitable sustainability tools. They can be usefully applied in the sustainability assessment and management process. Small subset of the most appropriate tools includes risk assessment, life cycle assessment, benefit cost analysis, ecosystem services valuation, integrated assessment models, sustainability impact assessment, and environmental justice tools.

Risk assessment is a tool widely used for characterizing the adverse human health and ecologic effects of exposures. Therefore, risk assessments can be classified into four major steps that include a hazard identification, close-response assessment, exposures assessment, and risk characterization. On the other hand, life cycle assessment is defined as a cradle to grave analysis or cradle to cradle of environment impacts of various products, to determine how changes in processes could lower the environmental impact, and to compare the environmental impacts of different products.

The main tool that is widely used for evaluating the net benefits of alternative decisions is the benefit cost analysis. It measures the change in welfare for each individual affected by policy choice. It is used to find a social net benefit and then rank the alternative. On the other hand, ecosystem services are goods and services that contribute to human well-being and the valuation measured in money terms can be used in benefit cost analysis to capture a more complete picture of the net benefit of alternative actions. On the other hand, sustainable impact assessment can be used to analyze the probable effects of a particular project on the three pillars of sustainability. It is used to develop integrated policies that take full account of the three sustainable development dimensions and long-term considerations of those policies.

Solid Waste Management

Waste is defined as any substances or objects that the holder discards or intends to discard. It can be classified into non-hazardous waste such as packaging waste and hazardous waste like chemical waste. Therefore, waste disposal should be seen as a last resort. Not only does waste disposal mean that valuable resources and energy are being thrown away but also biodegradable waste in landfill can emit methane. On the other hand, landfill space is becoming restricted.

Waste management has become a significant business issue for small businesses in recent years. All goods and products contain raw materials and energy. If they are discarded, we are effectively throwing away valuable natural resources. Waste disposal can also have adverse impacts on local air pollution and greenhouse gas emissions. Therefore, waste management can be defined as the collection, transport, processing, recycling and monitoring of waste materials that are produced by human. It is generally undertaken to reduce their effect on health, environment and carried out to recover resources from it.

The manufacturing strategy for environmentally kind products involves design process, which accounts for environmental impacts over the life of the products. Therefore, environmental improvements are related to manufacturing processes that are linked to reduction, reuse, recycling, and remanufacturing. However, eco-friendly comprises eco-design, eco-extraction, eco-manufacturing, eco-construction, eco-rehabilitation, eco-maintenance, eco-demolition, and socio-economic empowerment. Recently, sustainable eco-friendly road construction is increasingly receiving more attention worldwide. It is green our infrastructures of road constructing and reducing environmental impacts. It is also spurred by the increase in demand for eco-cities and eco-developments that are more environmentally friendly. Eco-friendly road construction can also be viewed as a response of stakeholders to the calls for sustainable development which arose from the growing awareness of the negative impact of road construction on our environments.

The quality of recycling wastes varies because of insufficient information on the properties of the manufacturing products, and lack of acknowledgement for using recycling materials as input material in new construction products, as well as lack of acknowledgement about the important elements and necessary actions for recycling the wastes.

The difficulties encountered in recycling are labor costs, lack of government awareness and support toward recycling, and limited real-life applications of recycled materials to allow for evaluation for their performance. The main benefits of recycling are reduction of material hauling and disposal costs and preservation of landfill capacity which lead to elongation of landfill design life and sometimes cheaper materials compared to virgin materials. Recycling helps in greening our infrastructures by conserving natural resources, decreasing energy use, reducing greenhouse gas emissions and air pollution, reducing the extraction of the virgin materials and minimizing their consumption, and environmental protection.

Different kinds of materials can be recycled in road construction such as fly ash, silica fume, ground granulated, blast furnace slag, reclaimed asphalt pavement, and plastic wastes such as polystyrene, polyethylene, and reclaimed concrete. Recycling waste materials can function as fine and coarse aggregates and supplementary cementing materials depending on the properties of the wastes intended to be optimized and the desired applications.

Eco-friendly road construction is one that is beneficial or non-harmful to the environment and is energy and resource efficient. To be eco-friendly, it must imbibe certain basic elements, namely eco-friendly, eco-extraction, eco-manufacturing, eco-construction, eco-rehabilitation, eco-maintenance, and eco-demolition. Utilization of waste materials will minimize negative impact on the environment and minimize the use of virgin materials.

An increasing in global plastic production and consumption due to increase or over population, development, and industrialization as well as lifestyle changes, the challenges posed by plastic wastes, which constitute of 25% of municipal solid waste. On the other hand, utilization of wastes from polyethylene constitutes 60% of plastic bottles. As a result, product packaging becomes the major contributor to environmental waste. Therefore, sustainable waste management will help to:

- Minimize waste.

- Reuse waste.

- Recycle waste for further use.

- Energy recovery.

- Disposal.

Sustainable Manufacturing Process

Sustainable manufacturing is defined as the creation of manufactured products that use processes that minimize negative environmental impacts, conserve energy and natural resources, are safe for employees, communities and consumers and are economically sound. Traditionally, manufacturing is defined as the process that is used to describe the physical transformation of materials or converting input materials into products. Sustainable production emphasizes a life cycle perspective in the manufacture, recycling, and disposal of goods and services, instead of the traditional focus on discrete activities. It encourages continuous to improved efficiency of using energy and resources. Therefore, sustainable production is defined as the creation of goods and services using processes and systems that are non-polluting, conserving energy and natural resources. Nowadays, sustainable manufacturing has evolved beyond the life cycle view and the key business has benefits from sustainable manufacturing, which includes financial performance, business excellence, and relationship with stakeholders as follows:

- Financial performance:

    ○ Increase sale.

- ◦ Improve efficiency and productivity by reducing resource use and waste.

- ◦ Reduce dependence and expensive or hazardous materials.

- Business excellence:

  - ◦ Stay ahead of regulations.

  - ◦ Win access to capital.

  - ◦ Gain strategic foresight.

- Relationship with stakeholders:

  - ◦ Enhance repetition.

  - ◦ Demonstrating green know-how and setting a positive example.

  - ◦ Improve employee's morale and retention.

  - ◦ Build better community relations.

Companies across the world face increased costs in materials, energy and compliance coupled with higher expectations of customers, investors and local communities because they throw away valuable natural resources as waste. Waste disposal has adverse impacts on the local air pollution and greenhouse gas emissions. Sustainable waste management is vital for:

- Saving valuable natural resources.

- Avoiding unnecessary emission of gas.

- Protective public health and natural ecosystems.

DOWNCYCLING

Downcycling is the term used to describe a recycled product that is not as structurally strong as the original product made from virgin materials.

Downcycled materials can therefore only be used to make a different type of product than the original. It is also possible to make the original product using recycled materials, provided there is a mix of recycled plastics with new, virgin materials.

Examples of Downcycling

When plastic bottles and materials are recycled by mechanical methods, the plastic gets weaker. However, downcycling makes it possible to still put the recycled materials to good use.

A common example of the downcycling process includes transforming plastic bottles into carpeting or fleece fibers and later turning fleece and carpeting materials into plastic lumber products.

Other downcycled uses for plastic bottles include:

- Car parts,
- Park benches,
- Drainage pipes,
- Railroad ties,
- Truck bed liners.

Polyethylene terephthalate (PET) plastic bottles can also be upcycled into goods of a higher quality, such as textiles.

Benefits of Downcycling

Downcycling helps protect the environment by eliminating waste and making new products out of something old that would otherwise end up in a landfill.

The downcycling process also offers benefits such as:

- Energy cost saving,

- Reduced pollution,

- Decreased manufacturing costs,

- Environmental protection.

Working Process of Downcycling

Downcycling happens all over the world in processing plants that accept plastic as well as other products like glass, cardboard, and paper.

Another benefit of downcycling from a manufacturer's point of view is that downcycled items do not have to be sorted. Unlike recycling, where a premium is put on sorting plastic bottles by material & color to ensure the clearest possible PCR for producing recycled bottles, common materials of any variety can be downcycled into a uniform, stable product.

The unsorted plastic containers are first broken down into plastic flakes. Manufacturers can then purchase the flakes for use in manufacturing processes for their respective industries. For instance, in the case of downcycling plastics into yarn or thread for carpet or garments, the flakes are pelletized, then extruded and spun into yarn. The same pelletization process is used in other applications including outdoor furniture, large containers (like garbage bins) and decking. The downcycled pellets can then be used for a variety of moulding applications similar to those used for virgin resins.

While plastic bottles and other materials lose their original structure and volume during the downcycling process, making it difficult if not impossible to recycle them again, it's still important to eliminate as much environmental waste as possible by repurposing into non-single use goods.

The Number of Times Plastics can be Recycled

Most plastics can be recycled into their original form only one or two times before they need to be downcycled into clothing or lumber products.

In some cases, the product resulting from downcycled plastics can't itself be recycled. For instance, if plastics are transformed into a fleece jacket, the jacket can't be downcycled like the original plastic bottle.

The Goal of Reduction

The rallying cry of environmentalists of 'Reduce, Reuse, Recycle' is certainly important, it's just as important to remember what order these terms are placed.

Ultimately, we should strive to reduce the use of materials as much as possible in our everyday life. Consider making minor lifestyle changes to reduce your impact on the environment, such as reusing that plastic water bottle before placing it in the recycling bin and reusing plastic grocery bags for future grocery trips.

Responsible manufacturers are also assuming their role in responsible recycling, often using recycled plastic bottles to make other products and working to reduce the amount and weight of their product packaging. They also strive to use highly-recyclable materials, like those sold through oberk.com, to minimize their impact on the planet.

UPCYCLING

The reduction of waste is an important goal that's shared by anyone who wants to help preserve the environment. Unfortunately, when the items that people use on a daily basis, such as glass jars, jeans, furniture, and tires, are no longer useful, they are typically thrown away. These are the items that ultimately fill up landfills and can damage the environment. While recycling is one alternative solution to throwing items in the garbage, it isn't the only option available. A fun and increasingly popular option is a process called upcycling. When a person upcycles old or undesired goods, they turn them into new and useful items that serve a different purpose. With imagination and the right know-how, nearly anyone can help the environment by upcycling.

Different ways to Upcycle

There are many different ways to upcycle depending on the object in question. A wine bottle, for example, can be painted or decorated and used as a vase for flowers or to hold vinegar or oil. Hang glass canning jars from hooks and place tea lights inside to create simple outdoor lanterns, or punch holes or designs into painted tin cans to create festive luminaries that can light up a front or backyard. Window shutters or an old door can be painted and used to create a unique headboard or table top. Other ways to upcycle goods include turning old-fashioned door knockers or silverware into handles for kitchen or bathroom cabinets and transforming baby cribs, suitcases, or old dresser drawers into bedding for one's pet. Even items such as cardboard boxes and tissue boxes can be upcycled. Adding a liner and wrapping twine or jute around cardboard boxes can turn them into attractive storage baskets, while an empty tissue box can be turned into a useful desktop caddy.

Impact of Upcycling in Environment

Because upcycling prevents materials from going to landfills, it saves valuable landfill space while also reducing the risk of toxic gases and other pollutants entering the atmosphere or poisoning the soil. People's unwanted items often end up in rivers and oceans as well, and upcycling can keep these materials from harming fish and other aquatic life. Certain items can be turned into clothing, handbags, or even jewelry. As a result, no manufacturing-related pollution is produced, no additional energy is wasted, and no new materials are consumed.

Applications

Art

Johann Dieter Wassmann (Jeff Wassmann),
Vorwarts.

The tradition of reusing found objects (*objet trouvé*) in mainstream art came of age sporadically through the 20th century, although it has long been a means of production in folk art. The Amish quilt, for example, came about through reapplication of salvaged fabric. Simon Rodia's Watts Towers in Los Angeles exemplifies upcycling of scrap metal, pottery and broken glass on a grand scale; it consists of 17 structures, the tallest reaching over 30 meters into the Watts skyline.

Intellectually, upcycling bears some resemblance to the ready-made art of Marcel Duchamp and the Dadaists. Duchamp's "Bicycle Wheel", a front wheel and fork attached to a common stool, is among the earliest of these works, while "Fountain", a common urinal purchased at a hardware store, is arguably his best-known work. Pablo Picasso's "Bull's Head", a sculpture made from a discarded bicycle saddle and handlebars, is the Spanish painter's sly nod to the Dadaists.

Throughout the mid-century, the artist Joseph Cornell fabricated collages and boxed assemblage works from old books, found objects and ephemera. Robert Rauschenberg collected trash and disused objects, first in Morocco and later on the streets of New York, to incorporate into his art works.

The idea of consciously raising the inherent value of recycled objects as a political statement, however, rather than presenting recycled objects as a reflection or outcome from the means of production, is largely a late 20th-century concept. Romuald Hazoumé, an artist from the West African Bénin, was heralded in 2007 for his use of discarded plastic gasoline and fuel canisters to resemble traditional African masks at Documenta 12 in Kassel, Germany. Hazoumé has said of these works, "I send back to the West that which belongs to them, that is to say, the refuse of consumer society that invades us every day."

Jeff Wassmann, an American artist who has lived in Australia for the past 25 years, uses items found on beaches and junk stores in his travels to create the early Modern works of a fictional German relative, Johann Dieter Wassmann. In *Vorwarts (Go Forward)* (pictured), Wassmann uses four simple objects to depict a vision of modern man on the precarious eave of the 20th century: an early optometry chart as background, a clock spring as eye, a 19th-century Chinese bone opium spoon from the Australian gold fields as nose and an upper set of dentures found on an Australian beach as mouth. Wassmann is unusual among artists in that he does not sell his work, rather they are presented as gifts; by not allowing these works to re-enter the consumer cycle, he averts the commodification of his end product.

Max Zorn is a Dutch tape artist who creates artwork from ordinary brown packaging tape and hangs pieces on street lamps as a new form of street art at night. By adding and subtracting layers of tape on acrylic glass with a surgical scalpel, the artwork can only be visible when light is placed behind it, mimicking the effects similar to stained glass window methods. His technique with pioneering upcycling with street art has been featured at Frei-Cycle 2013, the first design fair for recycling and upcycling in Freiburg, Germany.

Music

A prominent example is the Recycled Orchestra of Cateura in Paraguay,. The instruments of the orchestra are made from materials taken from the landfill of Asunción, whose name comes from the Cateura lagoon in the area. A limited part of its real history is narrated in the film "Landfill Harmonic", according to the version of Favio Chavez, ex-coordinator of "Sounds of the Earth".

In May 2016, an original member of the orchestra and two members of the parents' association, who helped build the current building of the Cateura music school; filed a complaint with the Office of the Prosecutor against the director's administration, for alleged lack of transparency. The complaint is backed by three senators from Paraguay.

Industry

Many industrial processes, like plastic and electronic fabrication, rely on the consumption of finite resources. Furthermore, the waste may have an environmental impact and can affect human health. Within this context, upcycling describes the use of available and future technologies to reduce waste and resource consumption by creating a product with a higher value from waste or byproduct streams. INDUSTRIAL UPCYCLING is a new complex methodology for waste prevention. In late 2015 on logistics conference LOGISTICS RIDE, Ostrava, Czech Republic, it was named for the first time INDUSTRY 5.0. It consists of tools and methods streamlining the production surplus from one process or factory to the next one. The originator of the complex methodology is Michael Rada (PROFILE), who started successful implementation within Czech republic industrial environment and expand to other locations.

In consumer electronics, the process of re-manufacturing or refurbishment of second-hand products can be seen as upcycling because of the reduced energy and material consumption in contrast to new manufacturing. The re-manufactured product has a higher value than disposing or downcycling it.

The use of Brewer's spent grain, a waste product of brewing processes, as a substrate in biogas processes eliminates the need for disposal and can generate significant profit to the overall brewing process. Depending on the substrate's price, a profit of approximately 20% of the operational costs is possible. In this process, the biogas plant acts as an "upcycler".

Libraries

Roses made from upcycled library books.

As libraries are becoming more than just bricks, mortar, and their collections, modern librarians are increasing their programming efforts. Upcycle programs or classes/clubs are becoming more and more popular, especially with the help of Pinterest. The Titusville Public Library has an upcycle class that started out by making roses.

Clothes

Designers have begun to use both industrial textile waste and existing clothing as the base material for creating new fashions. Upcycling has been known to use either pre-consumer or post-consumer waste or possibly a combination of the two. Pre-consumer waste is made while in the factory, such as fabric remnants left over from cutting out patterns. Post-consumer waste refers to the finished product when it's no longer useful to the owner, such as donated clothes.

Often, people practice linear economy where they are content to buy, use, then throw away. This system contributes to millions of kilos of textile waste being thrown away and makes fashion is the second-most polluting industry after oil. While most textiles produced are recyclable, around 85% end-up in landfills in the USA alone. Fast fashion companies are huge contributors to these problems as their whole purpose is to mass produce cheap clothing.

To live a sustainable life, clothing options opposite to the "throw away" attitude encouraged by fast fashion are needed. Upcycling can help with this, as it puts into practice a more circular economy model. A Circular Economy is where resources are used for as long as possible, getting the most value out of them while in use, then restored and repurposed when their use is over. Popularized by McDonough and Braungart, this has also been known as the *cradle-to-cradle* principle. This principle states a product should be designed either to have multiple life cycles or be biodegradable.

Not only does upcycling help environmentally, there are also economic benefits. By reusing materials that others may not find as desirable, there are huge savings that can be made on buying supplies for business or personal use. At schools or university, upcycling for art, craft, and design can be an easy and cheap way to prototype or finish a student project. There can also be an economic benefit with the final product. By definition, upcycling means increasing value to the original product and this can be done by changing/adding a personal spin to the design or improving material quality or fixing broken clothes.

Food

Billions of pounds of food are wasted every year around the world, but there are ways that people reuse food and find a way to upcycle. One common method is to feed it to animals because many animals, such as pigs, will eat all the scraps given. Food waste can be donated and restaurants can save all the food customers don't eat. Donations can also be made by contacting local agricultural extension offices to find out where to donate food waste and telling them how often and how much you can donate.

Another form of upcycling food would be to break it down and use it as energy. Engineers have found a way to break the food down into a reusable bio-fuel by pressure cooking it and then they are able to make methane out of the remains which can be used to produce electricity and heat.

When the food isn't used in those ways, another way is to just break it down and use it in compost, which will improve the soil. Many types of food waste, such as fruits, vegetables, egg shells, nuts, and nut shells, can be used in compost to enrich soil.

Design Processes

Tonnes of wastes are produced every day in our cities, and some educators are attempting to raise citizens' awareness, especially the youth. To redefine the concept of recycling previously confined to trash categorization, groups of young designers have attempted to transform "trash" into potentially marketable products such as backpacks made of waste plastic bags and area rugs created by reusing hides.

Hong Kong local inventor Gary Chan is actively involved in designing and making 'upcycling' bikes by making use of waste materials as parts for his bikes. He invented at least 8 bikes using wastes as a majority of the materials. Gary and his partners at Wheel Thing Makers regularly collect useful wastes such as leather skin from sofas, hardwood plates of wardrobes, or rubber tires from vehicle repair stores in the waste collection station on streets. Similar concepts evolve allover the world to tackle issue of rising excess of waste through creation of circular economies within the communities.

Potential Technologies

The worldwide plastic production was 280 million tons in 2011 and production levels are growing every year. Its haphazard disposal causes severe environmental damage such as the creation of the Great Pacific garbage patch. In order to solve this problem, the employment of modern technologies and processes to reuse the waste plastic as a cheap substrate is under research. The goal is to bring this material from the waste stream back into the mainstream by developing processes which will create an economic demand for them.

One approach in the field involves the conversion of waste plastics (like LDPE, PET, and HDPE) into paramagnetic, conducting microspheres or into carbon nanomaterials by applying high temperatures and chemical vapor deposition.

On a molecular level, the treatment of polymers like polypropylene or thermoplastics with electron beams (doses around 150 kGy) can increase material properties like bending strength and elasticity and provides an eco-friendly and sustainable way to upcycle them.

Active research is being carried out for the biotransformation upcycling of plastic waste (e.g., polyethylene terephthalate and polyurethane) into PHA bioplastic using bacteria.

PET could be converted into the biodegradable PHA by using a combination of temperature and microbial treatment. First it gets pyrolized at 450 °C and the resulting terephthalic acid is used as a substrate for microorganisms, which convert it finally into PHA. Similar to the aforementioned approach is the combination of nanomaterials like carbon nanotubes with powdered orange peel as a composite material. This might be used to remove synthetic dyes from wastewater.

Biotechnology companies have recently shifted focus towards the conversion of agricultural waste, or biomass, to different chemicals or commodities. One company in particular, BioTork, has signed an agreement with the State of Hawaii and the USDA to convert the unmarketable papayas in Hawaii into fish feed. As part of this Zero Waste Initiative put forth by the State of Hawaii, BioTork will upcycle the otherwise wasted biomass into a high quality, omega-rich fish feed.

WASTE HIERARCHY

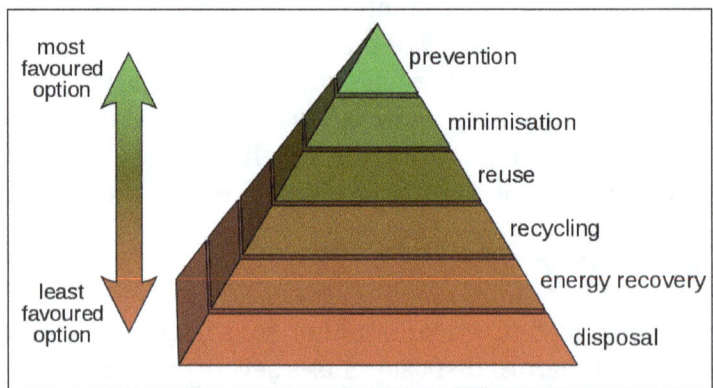

The waste hierarchy.

Waste hierarchy is a tool used in the evaluation of processes that protect the environment alongside resource and energy consumption from most favourable to least favourable actions. The hierarchy establishes preferred program priorities based on sustainability. To be sustainable, waste management cannot be solved only with technical end-of-pipe solutions and an integrated approach is necessary.

The three chasing arrows of the international recycling logo.
It is sometimes accompanied by the text "reduce, reuse and recycle".

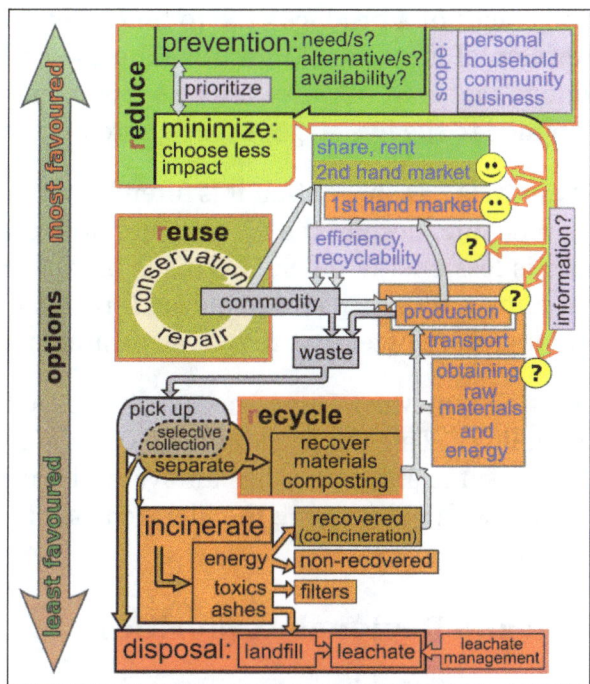

Enhanced version of waste hierarchy.

The waste management hierarchy indicates an order of preference for action to reduce and manage waste, and is usually presented diagrammatically in the form of a pyramid. The hierarchy captures the progression of a material or product through successive stages of waste management, and represents the latter part of the life-cycle for each product.

The aim of the waste hierarchy is to extract the maximum practical benefits from products and to generate the minimum amount of waste. The proper application of the waste hierarchy can have several benefits. It can help prevent emissions of greenhouse gases, reduces pollutants, save energy, conserves resources, create jobs and stimulate the development of green technologies.

Life-cycle Thinking

All products and services have environmental impacts, from the extraction of raw materials for production to manufacture, distribution, use and disposal. Following the waste hierarchy will generally lead to the most resource-efficient and environmentally sound choice but in some cases refining decisions within the hierarchy or departing from it can lead to better environmental outcomes.

Life cycle thinking and assessment can be used to support decision-making in the area of waste management and to identify the best environmental options. It can help policy makers understand the benefits and trade-offs they have to face when making decisions on waste management strategies. Life-cycle assessment provides an approach to ensure that the best outcome for the environment can be identified and put in place.

It involves looking at all stages of a product's life to find where improvements can be made to reduce environmental impacts and improve the use or reuse of resources. A key goal is to avoid actions that shift negative impacts from one stage to another. Life cycle thinking can be applied to the five stages of the waste management hierarchy.

For example, life-cycle analysis has shown that it is often better for the environment to replace an old washing machine, despite the waste generated, than to continue to use an older machine which is less energy-efficient. This is because a washing machine's greatest environmental impact is during its use phase. Buying an energy-efficient machine and using low- temperature detergent reduce environmental impacts.

The European Union Waste Framework Directive has introduced the concept of life-cycle thinking into waste policies. This duality approach gives a broader view of all environmental aspects and ensures any action has an overall benefit compared to other options. The actions to deal with waste along the hierarchy should be compatible with other environmental initiatives.

Challenges for Local and Regional Authorities

The task of implementing the waste hierarchy in waste management practices within a country may be delegated to the different levels of government (national, regional, local) and to other possible factors including industry, private companies and households. Local and regional authorities can be particularly challenged by the following issues when applying the waste hierarchy approach:

- A coherent waste management strategy must be set up.

- Separate collection and sorting systems for many different waste streams need to be established.

- Adequate treatment and disposal facilities must be established.

- An effective horizontal co-operation between local authorities and municipalities and a vertical co-operation between the different levels of government, local to regional and when beneficial, also at the national level need to established.

- Finding financing for the establishing or upgrading of expensive sustainable waste management infrastructure to address the needs of managing waste.

- A lack of data available on waste management strategies must be overcome and monitoring requirements must be met to implement the waste programs.

- The enforcement and control of business plans and practices be established and applied to maximize benefits to the environment and human health.

- A lack of administrative capacity at the regional and local level. The lack of finances, information, and technical expertise must be overcome for effective implementation and success of the waste management policies.

Source Reduction

Source reduction involves efforts to reduce hazardous waste and other materials by modifying industrial production. Source reduction methods involve changes in manufacturing technology, raw material inputs, and product formulation. At times, the term "pollution prevention" may refer to source reduction.

Another method of source reduction is to increase incentives for recycling. Many communities in the United States are implementing variable-rate pricing for waste disposal (also known as Pay As You Throw - PAYT) which has been effective in reducing the size of the municipal waste stream.

Source reduction is typically measured by efficiencies and cutbacks in waste. Toxics use reduction is a more controversial approach to source reduction that targets and measures reductions in the upstream use of toxic materials. Toxics use reduction emphasizes the more preventive aspects of source reduction but, due to its emphasis on toxic chemical inputs, has been opposed more vigorously by chemical manufacturers. Toxics use reduction programs have been set up by legislation in some states, e.g., Massachusetts, New Jersey, and Oregon. The 3 R's represent the 'Waste Hierarchy' which lists the best ways of managing waste from the most to the least desirable. Many of the things we currently throw away could be reused again with just a little thought and imagination.

WASTE MINIMISATION

Waste minimisation is a set of processes and practices intended to reduce the amount of waste produced. By reducing or eliminating the generation of harmful and persistent wastes, waste minimisation supports efforts to promote a more sustainable society. Waste minimisation involves redesigning products and processes and changing societal patterns of consumption and production.

The most environmentally resourceful, economically efficient, and cost effective way to manage waste often is to not have to address the problem in the first place. Managers see waste minimisation as a primary focus for most waste management strategies. Proper waste treatment and disposal can require a significant amount of time and resources; therefore, the benefits of waste minimisation can be considerable if carried out in an effective, safe and sustainable manner.

Traditional waste management focuses on processing waste after it is created, concentrating on re-use, recycling, and waste-to-energy conversion. Waste minimisation involves efforts to avoid creating the waste during manufacturing. To effectively implement waste minimisation the manager requires knowledge of the production process, cradle-to-grave analysis (the tracking of materials from their extraction to their return to earth) and details of the composition of the waste.

Waste hierarchy: Refusing, reducing, reusing, recycling
and composting allow to reduce waste.

The main sources of waste vary from country to country. In the UK, most waste comes from the construction and demolition of buildings, followed by mining and quarrying, industry and commerce. Household waste constitutes a relatively small proportion of all waste. Industrial waste is often tied to requirements in the supply chain. For example, a company handling a product may insist that it should be shipped using particular packing because it fits downstream needs.

Benefits

Waste minimisation can protect the environment and often turns out to have positive economic benefits. Waste minimisation can improve:

- Efficient production practices: Waste minimisation can achieve more output of product per unit of input of raw materials.

- Economic returns: More efficient use of products means reduced costs of purchasing new materials improving the financial performance of a company.

- Public image: The environmental profile of a company is an important part of its overall reputation and waste minimisation reflects a proactive movement towards environmental protection.

- Quality of products produced: New innovation and technological practices can reduce waste generation and improve the quality of the inputs in the production phase.

- Environmental responsibility: Minimising or eliminating waste generation makes it easier to meet targets of environmental regulations, policies, and standards. The environmental impact of waste will be reduced.

Industries

In industry, using more efficient manufacturing processes and better materials generally reduces the production of waste. The application of waste minimisation techniques has led to the development of innovative and commercially successful replacement products.

Waste minimisation efforts often require investment, which is usually compensated by the savings. However, waste reduction in one part of the production process may create waste production in another part.

Processes

- Reuse of scrap material: Scraps can be immediately re-incorporated at the beginning of the manufacturing line so that they do not become a waste product. Many industries routinely do this; for example, paper mills return any damaged rolls to the beginning of the production line, and in the manufacture of plastic items, off-cuts and scrap are re-incorporated into new products.

- Improved quality control and process monitoring: Steps can be taken to ensure that the number of reject batches is kept to a minimum. This is achieved by increasing the frequency of inspection and the number of points of inspection. For example, installing automated continuous monitoring equipment can help to identify production problems at an early stage.

- Waste exchanges: This is where the waste product of one process becomes the raw material for a second process. Waste exchanges represent another way of reducing waste disposal volumes for waste that cannot be eliminated.

- Ship to point of use: This involves making deliveries of incoming raw materials or components direct to the point where they are assembled or used in the manufacturing process to minimise handling and the use of protective wrappings or enclosures.

- Zero waste: This is a whole systems approach that aims to eliminate waste at the source and at all points down the supply chain, with the intention of producing no waste. It is a design philosophy which emphasizes waste prevention as opposed to end of pipe waste management. Since, globally speaking, waste as such, however minimal, can never be prevented (there will always be an end-of-life even for recycled products and materials), a related goal is pollution prevention.

Product Design

Waste minimisation and resource maximisation for manufactured products can most easily be done at the design stage. Reducing the number of components used in a product or making the product easier to take apart can make it easier to be repaired or recycled at the end of its useful life.

In some cases, it may be best not to minimise the volume of raw materials used to make a product, but instead reduce the volume or toxicity of the waste created at the end of a product's life, or the environmental impact of the product's use.

Fitting the Intended Use

In this strategy, products and packages are optimally designed to meet their intended use. This applies especially to packaging materials, which should only be as durable as necessary to serve their intended purpose. On the other hand, it could be more wasteful if food, which has consumed resources and energy in its production, is damaged and spoiled because of extreme measures to reduce the use of paper, metals, glass and plastics in its packaging.

Durability

Improving product durability, such as extending a vacuum cleaner's useful life to 15 years instead of 12, can reduce waste and usually much improves resource optimisation.

But in some cases it has a negative environmental impact. If a product is too durable, its replacement with more efficient technology is likely to be delayed. Therefore, extending an older machine's useful life may place a heavier burden on the environment than scrapping it, recycling its metal and buying a new model. Similarly, older vehicles consume more fuel and produce more emissions than their modern counterparts.

Most proponents of waste minimisation consider that the way forward may be to view any manufactured product at the end of its useful life as a resource for recycling and reuse rather than waste.

Making refillable glass bottles strong enough to withstand several journeys between the consumer and the bottling plant requires making them thicker and so heavier, which increases the resources required to transport them. Since transport has a large environmental impact, careful evaluation is required of the number of return journeys bottles make. If a refillable bottle is thrown away after being refilled only several times, the resources wasted may be greater than if the bottle had been designed for a single journey.

Many choices involve trade-offs of environmental impact, and often there is insufficient information to make informed decisions.

Retail

Various aspects of business practices affect waste, such as the use of disposable tableware in restaurants.

Reusable Shopping Bags

Reusable bags are a visible form of re-use, and some stores offer a "bag credit" for re-usable shopping bags, although at least one chain reversed its policy, claiming "it was just a temporary bonus". In contrast, one study suggests that a bag tax is a more effective incentive than a similar discount. (Of note, the before/after study compared a circumstance in which some stores offered a discount vs. a circumstance in which all stores applying the tax.) While there is a minor inconvenience involved, this may remedy itself, as reusable bags are generally more convenient for carrying groceries.

Households

Appropriate amounts and sizes can be chosen when purchasing goods; buying large containers of paint for a small decorating job or buying larger amounts of food than can be consumed create unnecessary waste. Also, if a pack or can is to be thrown away, any remaining contents must be removed before the container can be recycled.

Home composting, the practice of turning kitchen and garden waste into compost can be considered waste minimisation.

The resources that households use can be reduced considerably by using electricity thoughtfully (e.g. turning off lights and equipment when it is not needed) and by reducing the number of car journeys made. Individuals can reduce the amount of waste they create by buying fewer products and by buying products which last longer. Mending broken or worn items of clothing or equipment also contributes to minimising household waste. Individuals can minimise their water usage, and walk or cycle to their destination rather than using their car to save fuel and cut down emissions.

In a domestic situation, the potential for minimisation is often dictated by lifestyle. Some people may view it as wasteful to purchase new products solely to follow fashion trends when the older products are still usable. Adults working full-time have little free time, and so may have to purchase more convenient foods that require little preparation, or prefer disposable nappies if there is a baby in the family.

The amount of waste an individual produces is a small portion of all waste produced by society, and personal waste reduction can only make a small impact on overall waste volumes. Yet, influence on policy can be exerted in other areas. Increased consumer awareness of the impact and power of certain purchasing decisions allows industry and individuals to change the total resource consumption. Consumers can influence manufacturers and distributors by avoiding buying products that do not have eco-labelling, which is currently not mandatory, or choosing products that minimise the use of packaging. In the UK, PullApart combines both environmental and consumer packaging surveys, in a curbside packaging recycling classification system to minimise waste. Where reuse schemes are available, consumers can be proactive and use them.

Health-care Facilities

Health-care establishments are massive producers of waste. The major sources of health-care waste are: hospitals, laboratories and research centres, mortuary and autopsy centres, animal research and testing laboratories, blood banks and collection services, and nursing homes for the elderly.

Waste minimisation can offer many opportunities to these establishments to use fewer resources, be less wasteful and generate less hazardous waste. Good management and control practices among health-care facilities can have a significant effect on the reduction of waste generated each day.

Practices

There are many examples of more efficient practices that can encourage waste minimization in healthcare establishments and research facilities.

Source Reduction

- Purchasing reductions which ensures the selection of supplies that are less wasteful or less hazardous.

- The use of physical rather than chemical cleaning methods such as steam disinfection instead of chemical disinfection.

- Preventing the unnecessary wastage of products in nursing and cleaning activities.

Management and Control Measures at Hospital Level

- Centralized purchasing of hazardous chemicals.

- Monitoring the flow of chemicals within the health care facility from receipt as a raw material to disposal as a hazardous waste.

- The careful separation of waste matter to help minimize the quantities of hazardous waste and disposal.

Stock Management of Chemical and Pharmaceutical Products

- Frequent ordering of relatively small quantities rather than large quantities at one time.

- Using the oldest batch of a product first to avoid expiration dates and unnecessary waste.

- Using all the contents of a container containing hazardous waste.

- Checking the expiry date of all products at the time of delivery.

RECURSIVE RECYCLING

Recursive recycling is a technique where a function, in order to accomplish a task, calls itself with some part of the task or output from a previous step. In municipal solid waste and waste reclamation processing it is the process of extracting and converting materials from recycled materials derived from the previous step until all subsequent levels of output are extracted or used.

Level 1:

Solid waste or municipal solid waste can be treated, sanitized and separated under steam in a pressure vessel (waste autoclave). Following the processing under steam and removal of toxic materials via condensate filtering, usable recyclables are immediately extracted for reuse (plastics, ferrous metals, aluminum, glass, wood, etc.).

Level 2:

Organic materials from the original waste stream are converted to a fiber using steam at 60 psi and 160 °C. The converted organics (sanitary fiber) is size reduced by 85% and can be used to produce bio-fuels using acidic-hydrolysis or enzymatic-hydrolysis as ethanol or may be used as refuse derived fuel.

Level 3:

After the monosacrides are extracted for distillation, the remaining residue (used fiber) can be used as a feed stock for electricity production.

Level 4:

Finally, the non-toxic ash from the combusted fiber can be collected and used as a filler for preparation in super concrete and then reused in combination with similar materials (gravel, stones, pottery, glass) to form aggregate for construction materials.

In true recursive recycling and conservation processing the ability to divert all materials in the waste stream from landfill at greater than 99 percent is a concept based on outputs used to provide the next level of processing, reuse, conservation and market delivery of the derivatives.

Analyst Commentary

The concept of recursive recycling has been proved up in small scale facilities (thermal hydrolysis, plasma, etc.) but has not been widely accepted because of the financial impact it may have on existing protocols in waste management. One pilot facility operated commercially in Wales for approximately six years. However, the core equipment was moved to another location while the original facility was scheduled for retrofit. No further information about the facility's capacity or the equipment movement has been made available via open source release.

Since that pilot commercial facility stopped operating, the concept of recursive recycling has not met with as much success as originally anticipated by environmentalists and conservationists. There are a number of companies operating autoclaves with limited success across the globe but the full concept of waste treatment using thermal hydrolysis technology has not been fully realized because of several misconceptions in the autoclave and related steam treatment technologies. There is available engineering background to demonstrate successful testing that can be validated in physical production facilities but because of a lack of participation and general knowledge is a closely held secret, the application of technologies to achieve full recursive levels has not been accepted.

A number of companies are working on technology to support this concept. That may bring about change to conservation and recycling when associated advances prove out as successful. However, given the current state related areas of technology the growth to full-scale production appears to be limited because the technology and demonstrating it is available in larger capacities has not been demonstrated commercially.

RECOMMERCE

Recommerce or reverse commerce, refers to the process of selling previously owned, new or used products, mainly electronic devices or media such as books, through physical or online distribution channels to companies or consumers willing to repair, if necessary, and reuse, recycle or resell them afterwards.

In February 2005, in an interview for *The New York Times*, George F. Colony, chief executive of Forrester Research, was the first to introduce the term *recommerce* to answer a question about the increase in spending for technology after years of budget cuts in large corporations after the Dot Com Bubble burst: "There's a lot of shelf-life issues out there. People are a couple of releases behind. Older PCs. There is a move to really go back to - we call it 'recommerce'. Instead of 'ecommerce', it's 'recommerce'". He said.

The term later evolved, and today it is primarily used to describe business models centered on the purchase and resale of used goods by companies. Most recommerce businesses are focussed on consumer electronics, such as smartphones, tablets, and notebooks. Physical media, such as books, DVDs, and blue ray discs take a significant share of the recommerce industry as well.

While there was always an informal industry, such as garage sales and flea markets, to resell used goods, the creation of platforms, such as eBay or craigslist suddenly allowed private individuals to sell used goods of any kind much more efficiently.

In the recent years, beginning from the early 2000s, companies that professionalized the industry by offering professional buyback or trade-in schemes started to thrive: it became possible for consumers to sell their old smartphones, TVs, or computers when they purchased new ones to reduce the cost of the new device. This has been common practice with car sales for decades.

Companies, such as Gazelle, Rebag, and ThredUp started in the 2000s to offer purchasing and trade-in services completely separate from the purchase of new items and further spread the term.

Different Types of Recommerce

Informal Market

Consumers that sell used goods directly person to person (such as flea markets, garage sales or ad hoc) or via Marketplaces such as Amazon, eBay. Hereby some platforms such as eBay may hedge the risk of the payment for the consumer by providing payment tools such as PayPal or just offer the possibility to market the product such as craigslist.

Trade-in and Recommerce Services

An increasing amount of transactions occur directly via specialised service companies that purchase used goods, refurbish them if necessary and resell them subsequently. Such platforms often provide initial indications of the final purchase price for the good.

Most platforms assist the user during the transaction by offering following services:

- An indication of the final purchase price to the owner of the product sold (often the final price varies as the consumer cannot verify to the full extend all determining factors of the product sold such as the quality of the battery of a smartphone during the process).
- By organizing the logistical return of the product.
- By controlling the product's condition in a specialized workshop.
- By recycling the good if it can't be used anymore.

This kind of resale allows sellers to overcome some marketplaces drawbacks, by providing a means of simplified acquisition, and immediate and sometimes guarantied value.

Buy Back and Trade-in Offers by Vendors of New Products

Especially in the electronics sector the purchase and buyback of consumer electronics became very common in recent years. By today all major MNO offer Trade-In solutions combined or detached from the purchase of a new phone. Most of this services are offered by 3rd party refurbishing companies specialised in used electronics.

Mobile Operator	Offers Trade-In/Buy Back
AT&T	Yes, all major brands
T-Mobile	Yes, all major brands
Verizon	Yes, all major brands
Sprint	Yes, all major brands

Types of Purchased Products

Examples of the main assets of companies acquired by recommerce include:

- Consumer non-durables: disposable razors, jeans, corks, pantyhose, eyeglasses, watches.

- Cultural goods: books, CDs, DVDs.

- Jewelry: gold, silver.

- Technological devices: cellphones, smartphones, tablet computers, TVs, video game consoles, GPS devices, cameras, video cameras.

- Clothes & unwanted fashion items and accessories; handbags, small goods.

- Over the counter (OTC) medical supplies, particularly Diabetes testing supplies such as glucose test strips and lancets.

Many ecommerce services have introduced recommerce solutions, including distributors, online retailers, and chain retailers.

The various Marketing Positions Recommerce

Multiple types of recommerce services are available:

- Recommerce, used as a method of funding, which can compensate the seller in cash or with a voucher.

- Solidarity recommerce, the return of products by offering its holders an opportunity to share or redistribute the residual value with a non-profit organization or a social cause (e.g., micro-credit, or insertion).

- Ecological recommerce, the recycling or proper disposal of products with strong polluting capacity (providing repair or recycling regulatory compliance).

Positive Impact of Recommerce

Environment

Environmental reports by electronics manufactures show that the majority of natural resources for the production of such products are consumed during manufacturing and first transport of the product and not during the use of a product. In many cases the reuse of such a good is significantly more beneficial than the pure recycling as eventual logistics and energy consumptions during the recycling don't occur and a used product can be resold instead of a new product being produced. The reuse of a product is an effective means of reducing products' environmental footprint.

Smartphone	Production	Logistics	Use	Recycling	Total Emissions
iPhone 6	85%	3%	11%	1%	95 kg CO_2e

Consumer Purchase Power

The product holder increases its purchasing power to acquire a new product by selling back its old one.

The Development Factors

Several factors have greatly accelerated the development of recommerce in developed countries, including:

- The demand for solutions enabling consumers to separate themselves ethically from their products.

- The ease of use of recommerce services, and more importantly.

- The preponderance of takeover bids, which handsomely compensate the owners of recommerce services.

In France, the rise of recommerce is partly supported by the "Grenelle II" Law, which states that when they are sold under the brand name of a single dealer, it must "provide or contribute to the collection, removal and treatment of electrical and electronic equipment waste instead of the person who manufactures, imports or brings in the domestic market this equipment regardless of the selling technique, including distance selling and electronic sales".

Issues of Recommerce

Recommerce requires a special organization of many functions, such as: logistics management, information systems, customer relations, price control and treatment of the product in the shop, promotion, retention, and resale. Functional products recovered via recommerce solutions are usually put back on the market by the recommercer. Moreover, when this product exceeds local demand, recommercers sometimes turn to foreign markets to sell the products they have purchased. Thus, the recommercer sells some of these used functional products in emerging markets where access to technology and accelerating economic development are reserved for some part of the population.

RECYCLING SYMBOL

The Universal Recycling Symbol, here rendered with a black outline and green fill. Both filled and outline versions of the symbol are in use.

The universal recycling symbolis internationally recognized for recycling activity.

Worldwide attention to environmental issues led to the first Earth Day in 1970. Container Corporation of America, a large producer of recycled paperboard, sponsored a contest for art and design students at high schools and colleges across the country to raise awareness of environmental issues. It was won by Gary Anderson, then a 23-year-old college student at the University of Southern California, whose entry was the image now known as the universal recycling symbol. The symbol is not trademarked and is in the public domain. The public-domain status of the symbol has been challenged, but this challenge was unsuccessful owing to the wide use of the symbol. However, the universal symbol may have been inspired by similar existing recycling symbols, such as one featuring two arrows chasing each other in a circle that Volkswagen stamped in the early 1960s into some automobile parts it remanufactured.

Variants

The recycling symbol is in the public domain, and is not a trademark. The Container Corporation of America originally applied for a trademark on the design, but the application was challenged, and the corporation decided to abandon the claim. As such, anyone may use or modify the recycling symbol, royalty-free.

Though use of the symbol is regulated by law in some countries, countless variants of it exist worldwide. Anderson's original proposal had the arrows form a triangle standing on its tip—upside down compared with the versions most commonly seen today—but the CCA, in adopting Anderson's design, rotated it 60° to stand on its base instead.

Both Anderson's proposal and CCA's designs form a Möbius strip with *one* half-twist by having two of the arrows fold over each other, and one fold under, thereby canceling out one of the other folds. However, most variants of the symbol used today have all the arrows folding over themselves, producing a Möbius strip with *three* half-twists. Existing single half-twist variants of the logo do not generally agree on which of the arrows is the one to fold underneath. The logo is usually displayed with the arrows circulating clockwise, but the underlying Möbius strip exists in two topologically distinct mirror-image forms of opposite handedness.

The American Paper Institute originally promoted four different variants of the recycling symbol for different purposes. The plain Möbius loop, either white with an outline or solid black, was to be used to indicate that a product was *recyclable*. The other two variants had the Möbius loop inside a circle—either white on black or black on white—and were meant for products *made of recycled materials*, with the white-on-black version to be used for 100% recycled fiber, and the black-on-white version for products containing both recycled and unrecycled fiber. For example, a paper envelope might have both the first and last of these four symbols, to indicate that it was recyclable, and made from both recycled and unrecycled fibers.

In addition to the resin identification codes 1–7 in the triangular recycling symbol, Unicode lists the following recycling symbols:

- U+2672 ♲ universal recycling symbol.

- U+267A ♺ recycling symbol for generic materials.

- U+267B ♻ black universal recycling symbol.

- U+267C ♼ recycled paper symbol (indicates product contains recycled paper).

- U+267D ♽ partially-recycled paper symbol (indicates product contains partially recycled paper).

- U+267E ♾ permanent paper sign (e.g. for acid-free paper).

An ISO/IEC working group has researched and documented some of the variations of the recycling logo currently in use, and has made recommendations for adding some more of them to the Unicode standard.

With the rapid expansion of materials converted to printer filament for 3-D printing using recyclebot technology, a large expansion of resin identification codes has been proposed.

Resin Identification Code

In 1988, the American Society of the Plastics Industry (SPI) developed the resin identification code which is used to indicate the predominant plastic material used in the manufacture of the product or packaging. Their purpose is to assist recyclers with sorting the collected materials, but they do not necessarily mean that the product/packaging can be recycled either through domestic curbside collection or industrial collections. The SPI symbols are loosely based on the Möbius loop symbol, but feature simpler bent (rather than folded over) arrows that can be embossed on plastic surfaces without loss of detail. The arrows are formed into a flat, two-dimensional triangle rather than the pseudo-three-dimensional triangle used in the original recycling logo.

The different resin identification codes can be represented by Unicode icons ♳ (U+2673), ♴ (U+2674), ♵ (U+2675), ♶ (U+2676), ♷ (U+2677), ♸ (U+2678), ♹ (U+2679), and ♺ (U+267A).

Recycling codes extend these numbers above 7 to include various non-plastic materials, including metals, glass, paper, and batteries of various types.

Other Variants

Taiwan's recycling symbol features the use of negative
space to also create arrows pointing outward.

An infinity sign (∞) inside a circle, represents the permanent paper symbol, used in packaging and publishing to signify the use of durable acid-free paper. In some ways, this logo expresses the *opposite* intention from the recycle logo, in that the acid-free paper is intended to last indefinitely, rather than being recycled. Nevertheless, acid-free paper does not usually contain toxic materials (although certain inks do), so it is easily recycled or composted.

A satirical version of the classic recycling logo also exists, in which the three arrows are twisted from a circular pattern to pointing radially outward, thus symbolizing wasteful one-time usage rather than environmentally friendly recycling. This message is reinforced by the circular inscription, "This project was environmentally unfriendly", surrounding the modified logo. The satirical logo appears in the 1998 catalog of an installation art work in Bayonne, New Jersey, in which the artist Steven Pippin modified a row of glass-doored washing machines in a laundromat to operate as giant cameras. The cameras were used to take sequential photographs in the manner of pioneering stop motion photographer Eadweard Muybridge. The front-loading washing machines were then used to develop and process the 24 inch (61 cm) diameter circular film negatives.

RECYCLING CODES

Recycling codes on products.

Recycling codes are used to identify the material from which an item is made, to facilitate easier recycling or other reprocessing. Having a recycling code, the chasing arrows logo or a resin code on an item is not an automatic indicator that a material is recyclable but rather an explanation of what the item is. Such symbols have been defined for batteries, biomatter/organic material, glass, metals, paper, and plastics.[*citation needed*] Various countries have adopted different codes. For example, the table below shows the polymer resin codes (plastic) for a country. In the United States there are fewer, as ABS is grouped in with others in group 7. Other countries have a more granular recycling code system. For example, China's polymer identification system has seven different classifications of plastic, five different symbols for post-consumer paths, and 140 identification codes. The lack of codes in some countries has encouraged those who can fabricate their own plastic products, such as RepRap and other prosumer 3-D printer users, to adopt a voluntary recycling code based on the more comprehensive Chinese system.

Plastic Recycling Codes

Every town and city has different recycling programs, so you'll often have to check your location's rules to find out exactly what you can recycle. Plus, "there are times when your recycling program may change what it collects," Even if there isn't a way for your town to recycle a certain material, he says there's still a chance they might collect it anyways and either store it or dispose of it.

Plastic Recycling Symbol: PET or PETE

PET or PETE (polyethylene terephthalate) is the most common plastic for single-use bottled beverages, because it's inexpensive, lightweight, and easy to recycle. It poses low risk of leaching breakdown products. Its recycling rates remain relatively low (around 20%), even though the material is in high demand by manufacturers.

Found in: Soft drinks, water, ketchup, and beer bottles; mouthwash bottles; peanut butter containers; salad dressing and vegetable oil containers.

How to recycle it: PET or PETE can be picked up through most curbside recycling programs as long as it's been emptied and rinsed of any food. When it comes to caps, our environmental pros say it's probably better to dispose of them in the trash (since they're usually made of a different type of plastic), unless your town explicitly says you can throw them in the recycle bin. There's no need to remove bottle labels because the recycling process separates them.

Recycled into: Polar fleece, fiber, tote bags, furniture, carpet, paneling, straps, bottles and food containers (as long as the plastic being recycled meets purity standards and doesn't have hazardous contaminants).

PET
(POLYETHYLENE
TEREPHTHALATE)

Plastic Recycling Symbol: HDPE

HDPE (high density polyethylene) is a versatile plastic with many uses, especially when it comes to packaging. It carries low risk of leaching and is readily recyclable into many types of goods.

Found in: Milk jugs; juice bottles; bleach, detergent, and other household cleaner bottles; shampoo bottles; some trash and shopping bags; motor oil bottles; butter and yogurt tubs; cereal box liners.

How to recycle it: HDPE can often be picked up through most curbside recycling programs, although some allow only containers with necks. Flimsy plastics (like grocery bags and plastic wrap) usually can't be recycled, but some stores will collect and recycle them.

Recycled into: Laundry detergent bottles, oil bottles, pens, recycling containers, floor tile, drainage pipe, lumber, benches, doghouses, picnic tables, fencing, shampoo bottles.

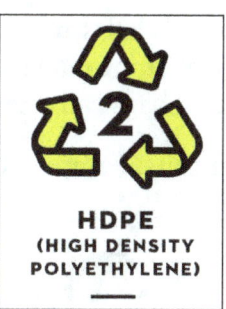

HDPE
(HIGH DENSITY
POLYETHYLENE)

Plastic Recycling Symbol: PVC or V

PVC (polyvinyl chloride) and V (vinyl) is tough and weathers well, so it's commonly used for things like piping and siding. PVC is also cheap, so it's found in plenty of products and packaging. Because chlorine is part of PVC, it can result in the release of highly dangerous dioxins during manufacturing. Remember to never burn PVC, because it releases toxins.

Found in: Shampoo and cooking oil bottles, blister packaging, wire jacketing, siding, windows, piping.

How to recycle it: PVC and V can rarely be recycled, but it's accepted by some plastic lumber makers. If you need to dispose of either material, ask your local waste management to see if you should put it in the trash or drop it off at a collection center.

Recycled into: Decks, paneling, mud-flaps, roadway gutters, flooring, cables, speed bumps, mats.

PVC
(POLYVINYL
CHLORIDE)

Plastic Recycling Symbol: LDPE

LDPE (low density polyethylene) is a flexible plastic with many applications. Historically, it hasn't been accepted through most American recycling programs, but more and more communities are starting to accept it.

Found in: Squeezable bottles; bread, frozen food, dry cleaning, and shopping bags; tote bags; furniture.

How to recycle it: LDPE is not often recycled through curbside programs, but some communities might accept it. That means anything made with LDPE (like toothpaste tubes) can be thrown in the trash. Plastic shopping bags can often be returned to stores for recycling.

Recycled into: Trash can liners and cans, compost bins, shipping envelopes, paneling, lumber, landscaping ties, floor tile.

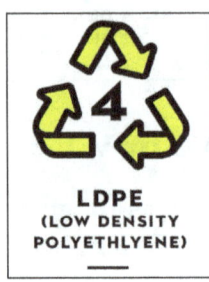

Plastic Recycling Symbols: PP

PP (polypropylene) has a high melting point, so it's often chosen for containers that will hold hot liquid. It's gradually becoming more accepted by recyclers.

Found in: Some yogurt containers, syrup and medicine bottles, caps, straws.

How to recycle it: PP can be recycled through some curbside programs, just don't forget to make sure there's no food left inside. It's best to throw loose caps into the garbage since they easily slip through screens during recycling and end up as trash anyways.

Recycled into: Signal lights, battery cables, brooms, brushes, auto battery cases, ice scrapers, landscape borders, bicycle racks, rakes, bins, pallets, trays.

Plastic Recycling Symbol: PS

PS (polystyrene) can be made into rigid or foam products — in the latter case it is popularly known as the trademark Styrofoam. Styrene monomer (a type of molecule) can leach into foods and is a possible human carcinogen, while styrene oxide is classified as a probable carcinogen. The material was long on environmentalists' hit lists for dispersing widely across the landscape, and for being notoriously difficult to recycle. Most places still don't accept it in foam forms because it's 98% air.

Disposable plates and cups, meat trays, egg cartons, carry-out containers, aspir... bottles, compact disc cases.

How to recycle it: Not many curbside recycling programs accept PS in the form of rigid plastics (and many manufacturers have switched to using PET instead). Since foam products tend to break apart into smaller pieces, you should place them in a bag, squeeze out the air, and tie it up before putting it in the trash to prevent pellets from dispersing.

Recycled into: Insulation, light switch plates, egg cartons, vents, rulers, foam packing, carry-out containers.

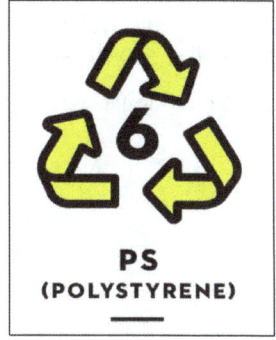

Plastic Recycling Symbol: Miscellaneous

A wide variety of plastic resins that don't fit into the previous categories are lumped into this one. Polycarbonate is number seven plastic, and it's the hard plastic that has worried parents after studies have shown it as a hormone disruptor. PLA (polylactic acid), which is made from plants and is carbon neutral, also falls into this category.

Recycled into: Plastic lumber and custom-made products.

Found in: Three- and five-gallon water bottles, bullet-proof materials, sunglasses, DVDs, iPod and computer cases, signs and displays, certain food containers, nylon.

How to recycle it: These other plastics are traditionally not recycled, so don't expect your local provider to accept them. The best option is to consult your municipality's website for specific instructions.

References

- Emily J. Hunt, Chenlong Zhang, Nick Anzalone, Joshua M. Pearce, Polymer recycling codes for distributed manufacturing with 3-D printers, Resources, Conservation and Recycling, 97, pp. 24-30 (2015). DOI:10.1016/j.resconrec.2015.02.004

- Recyclate, engineering: sciencedirect.com, Retrieved 21 April, 2020

- Sustainable-solid-waste-recycling, skills-development-for-sustainable-manufacturing: intechopen.com, Retrieved 17 March, 2020

- "Rules Governing Use of Recycling Logo". Recycling Expert. recyclingexpert.co.uk. Retrieved 2011-05-16

- Why-is-recycling-important: conserve-energy-future.com, Retrieved 9 June, 2020

- Claudia E. Henninger; Panayiota J. Alevizou; Helen Goworek; Daniella Ryding, eds. (2017). Sustainability in Fashion. doi:10.1007/978-3-319-51253-2. ISBN 978-3-319-51252-5

- Recycling-symbols-plastics-460321: goodhousekeeping.com, Retrieved 7 April, 2020

2

Recycling of Materials

Recycling can be carried out on a variety of materials such as concrete, glass, plastic, timber, cotton, gypsum, paper, paint, vegetable, timber and metals including aluminium, copper, iron and steel, etc. The topics elaborated in this chapter will help in gaining a better perspective of the recycling of different materials.

FERROUS METALS RECYCLING

Ferrous metals are pure iron or an alloy that contains iron. The most common ferrous metal alloy is steel. Ferrous metals have small amounts of other metals or elements added, to give the required properties. These metals are magnetic and provide very little resistance to corrosion.

All commercial forms of iron and steel contain carbon, which has become an integral part of the metallurgy of iron and steel. The demand for ferrous metals is strong, thus scrap metals are also highly sought after. Most recycling companies pay for scrap metals.

Types and Features of Ferrous Metals

There are many variations of ferrous metals available in the market today. Some of the key types and their features are listed below:

- Cast iron: Hard, brittle, strong, cheap, self-lubricating.

- Mild steel: Tough, high tensile strength, ductile. Because of low carbon content it can not be hardened and tempered. It must be case hardened.

- High carbon steel: The hardest of the carbon steels. Less ductile, tough and malleable.

- Stainless steel: Corrosion resistant.

Other available types are construction steel, free cutting steel, high strain steel high temperature steel, low temperature steel, and spring steel.

Manufacturing Process of Ferrous Metals

Ferrous metals have to be extracted from iron ore. The source for iron ore is the earth's crust, which contains metals and metal compounds such as iron oxide. However, the ore is often mixed with other substances. In order to optimize the usage of the metal, it has to be extracted from the mixture.

The method primarily used to extract metals from the ore depends on their reactivity. In the case of iron, a less-reactive metal, it can be extracted by reduction with carbon or carbon monoxide. Iron is then extracted from iron ore in a huge container called a blast furnace. Oxygen must be removed from the iron oxide to leave the iron behind.

Applications of Virgin Material Ferrous Metals

Ferrous metals are used in numerous applications. The following are key uses:

- Girders, plates, nuts and bolts.

- Heavy crushing machinery.

- Car cylinder blocks, machine tool parts, vices, brake drums, machine handle and gear wheels, and plumbing fitments.

- Gears, shafts, engine parts.

- Cutting tools for lathes.

- Kitchen draining boards, pipes, cutlery, aircraft.

- Chisels, files, lathe tools, hammers, drills, taps and dies.

- Metal ropes, springs, wire, garden tools.

Environmental Impacts of Ferrous Metals

Scrap ferrous metals such as car bodies and old farm machinery contain many toxic chemicals including oil, petrol and diesel, battery acids, transmission and brake fluids plus radiator coolant that can leach into the environment when dumped in landfills or other illegal places. These scrap metals also create safety and fire hazards. The sharp jagged edges of the rusting metal might injure people and wildlife, while the chemicals could cause bush fires.

Illegally dumped scrap metal is expensive to clear by the local government. These costs will include collection and disposal of material, increased enforcement and surveillance, prosecution, education and awareness programs.

Recycling Process

The use of scrap ferrous metal has become a core part of most steelmaking companies today as it helps improves the company's economic viability and whilst decreasing environmental impact. Another reason is that in comparison to ore extraction, using scrap ferrous metals reduces energy consumption, CO_2 emissions, water consumption, and air pollution.

Recycling of ferrous metals helps to reduce the quantities of solid waste deposited in landfills, which have become more expensive. As these metals are magnetic, scrap can be collected using a magnet and sent to a recycling unit. Ferrous metal-based scrap products can be recycled by remelting, recasting, and redrawing processes completely within a steel mill. These processes are far cheaper than producing new metal from the basic ore. Manufacturers produce their own coke. The by-products from the coke oven include several organic compounds, ammonia, and hydrogen sulfide. All these can be sold to various consumers.

One of the largest sources of scrap steel is reprocessing old automobile bodies. The automobile body is crushed and flattened, and then is shredded into small pieces by hammer mills. Ferrous metals are separated from the shredder residue using powerful magnets while other materials are sorted manually or by using high-pressure air flows and liquid floating systems.

Applications of Recycled Ferrous Metals

Studies have shown that recycled steel has the same strength as new steel. Today most appliances contain about 75% recycled steel.

NON-FERROUS RECYCLING

Nonferrous scrap metal is scrap metal other than iron and steel. Nonferrous scrap metal includes aluminum, which includes foil and cans, copper, lead, zinc, nickel, titanium, cobalt, chromium, and precious metals. While the volume of non-ferrous scrap metal is less then ferrous scrap, it is more valuable by the pound then ferrous scrap metal.

Scrap metal processors handle vast quantities of non-ferrous scrap metal yearly. It is estimated that scrap yards process approximately:

- 1.5 million tons of copper scrap annually.

- 2.5 million tons of aluminum scrap annually.

- 1.3 million tons of lead scrap annually.

- 300,000 tons of zinc scrap annually.

Furthermore all metals, including both ferrous and non-ferrous can be recycled indefinitely without loosing any of their properties, an attractive feature to many scrap metal processors. The ever-increasing value of non-ferrous metals from aluminum to exotic metal shavings has resulted in the need for specialized baling equipment. Harmony Enterprises can provide the appropriate solution for any size enterprise from small metal-fabricating firms to large-scale industrial production facilities.

Benefits to Non-ferrous Metal Recycling

The use of both ferrous and non-ferrous metals has continued to increase over the years. The amount of aluminum in production alone is astonishing. It is estimated that world primary production of aluminum is around 24 million tons on average per year, of which Australia is the largest producer. With metal production in general on the rise, and the valuable resources necessary for their production dwindling, it is more important then ever to find ways to find further value out of scrap metal reuse.

Whereas plastic packaging may cause difficulties in the recycling process due to its disparate nature and need to separate it into its different types – metal packaging recycling is simpler. As the ferrous and non-ferrous components can be separated using magnets one large problem of identification is solved.

Additionally the recovery of non-ferrous metals for recycling provides both environmental and economic benefits:

- Non-Ferrous Metal Recycling frees space.

- Non-Ferrous Metal Recycling creates a safer work environment.

- Non-Ferrous Metal Recycling results in less pollution, greater energy savings, and is overall more environmentally friendly.

- Non-Ferrous Metal Recycling creates additional employment opportunities.

- Non-Ferrous Metal Recycling provides additional materials for reuse in manufacturing.

- Non-Ferrous Metal Recycling is a source of revenue.

- Non-Ferrous Metal Recycling encourages the development of additional markets.

Non-ferrous Metals

Aluminum

Produced from bauxite, aluminum is a clay-like ore that is rich in aluminum compounds. Aluminum is only found as a compound called alumina, which is a hard material consisting of aluminum combined with oxygen. In order to free the aluminum, the alumina has to be stripped of its oxygen, and dissolved in a molten salt at a reduction plant after which a powerful electric current is run though the liquid to separate the aluminum from the oxygen. This process uses large quantities of energy. One of the most cost-effective materials to recycle, aluminum recycling requires a lot less energy then aluminum production. In fact recycling aluminum requires only 5% of the energy and produces only 5% of the CO_2 emissions as compared with primary production, all while reducing the amount of waste going to landfill. An additional added bonus is that aluminum can be recycled indefinitely, as reprocessing does not damage its structure.

Other Metals

All other non-ferrous metals, though often present in smaller quantities can be recycled as well. These metals which include nickel, copper, silver, gold, lead, and brass are heavily relied on by specific industries, such as the Electronic and Technology Industry. Due to their recognized value, smaller quantities of these metals are in circulation and their ability to be recycled is often neglected when people dispose of these items.

Baling Non-ferrous Metals

The non-ferrous metal recycling process begins by gathering bulky non-ferrous metals, such as clippings from industrial manufacturing process, aluminum beverage cans, and obsolete scrap, and baling these materials into various sized blocks of bales. Front-end loaders and conveyors, along with other waste handling equipment are utilized to feed the bulky materials into the baler.

There are wide variety of recycling baler types for aluminum scrap and can recycling. Some balers, such as the T60XDRC are designed to compress bulky materials at a very high pressure, into dense, uniformly sized bales, which are able to be efficiently remelted by our customers. Other balers like the M60STD are designed to compress bulky material at lower pressure into bigger, less dense, uniformly sized bales, able to be more efficiently stored and transported to consumers who require certain material delivered in this form.

Actual bale weights of non-ferrous metal vary greatly depending upon the type of material being baled, and actual size and configurations of the non-ferrous baler itself. Below you will find some common loose and baled weights of non-ferrous metal and various other materials.

Type of Material	Loose	Baled
Cardboard	50 – 100 lbs/cy	600 – 1100 lbs / cy
PET (Soda bottles, food packaging etc)	30 – 40 lbs / cy	200–500 lbs / cy
HDPE (Milk Jugs, Detergent Containers etc)	22 – 25 lbs / cy	200 – 500 lbs / cy
Aluminum Cans	50 – 75 lbs / cy	150–500 lbs / cy
Steel Cans	150 – 175 lbs / cy	500 – 1,000 lbs / cy
Paper	500– 600 lbs / cy	1,000 – 1,200 lbs / cy
Newspaper	350 – 500 lbs / cy	750 – 1,000 lbs / cy

Methods of Non-ferrous Metal Recycling

Recycling non-ferrous metals can be a problematic task as they will not always come in their pure form. Often times, they are found in all sorts of liquid and solid mixtures from which they need to be extracted and purified before further use. Three methods used today for non-ferrous metal recycling are electrowinning, precipitation, and non-ferrous sensors.

Electrowinning

Electrowinning, which is also known as electroextraction, is, on the surface of it at least, a relatively simple process of extracting dissolved metals from their dissolved states using electricity. In case of non-ferrous metal extraction for the purpose of recycling, the process generally goes as follows. First, the material, which can be any form of waste such solid materials from landfills or different types of solid mixtures containing non-ferrous metals, is put into a liquid solution where it is dissolved into a liquid state through the process known as leaching, the end-result of this process is called a leach-ate or leach solution. Then, using an anode and cathode – which are electrodes through which the current flows and which are submerged into the solution – an electric current is passed through the leach solution which then causes the metals to be (chemically) reduced resulting in them forming a thin even layer across the surface of the submerged cathode. This way, the non-ferrous metals, such as copper, tin, nickel, or silver are recovered and made readily available for further reuse.

Precipitation

The second processing method for non-ferrous metals is precipitation. It is also the most widely used method for metal recovery from aqueous solutions. Precipitation can also be used for wastewater treatment; a process in which metals are recovered from aqueous waste solutions.

This method includes two metal removal sub-methods called co-precipitation and adsorption. So as not to go into too much technical details, we will mainly address the basic method of precipitation without going too much into other of its aspects. Precipitation is the process of forming of an insoluble solid material from what originally was an aqueous solution typically involving pH adjustment or addition of another chemical species.

The end-result is called the "precipitate" while the chemical that causes this is called the "precipitant." The most commonly used precipitants are sodium and calcium hydroxides or oxides which are used to increase the pH resulting in insoluble metal hydroxides.

Metal Sensors

Finally, non-ferrous metal sensors are becoming widely used in sorting and extracting non-ferrous metals from scrap, most of which originated from end-of-life-vehicles or from e-waste. For example, sensors are used for detection and extraction of specific non-ferrous metals from Zobra.

According to the Institute for Scrap Recycling Industries in the United States, Zobra is defined as a mixture of shredded non-ferrous scrap metals primarily consisting of aluminum but also containing copper, lead, brass, zinc, tin, nickel and copper in any of their forms. Given the fact that we have already seen that using recycled aluminum can result in big savings, and the demand for other non-ferrous metals found in Zobra, the commercial potential of this mixture as well as the economic significance of extracting and sorting out these elements from the mixture becomes rather obvious. Typically, these metals would be sorted either manually or by using 'sink-float' gravimetric treatments.

However, both methods are largely unreliable as the first method relies on human intervention and observation, while the second method relies on the density of said materials. The problem arises because some of the non-ferrous metals are of similar densities so they will not be separated from each other using gravimetric techniques. Using sensors, including for example X-ray transmission technology which can target different materials based on their atomic density, is significantly more reliable and almost completely removes arbitrariness enabling the enhanced separation of the scrap materials acquired.

For example, in the case of finely cut copper containing wires, which are often accompanied by traces of brass or stainless steel, using sensor-based technology makes it possible to detect and remove copper particles smaller than 1mm in size from the mixture ensuring purity of greater than 99%; a rate that could never be matched by most sophisticated sink-float mechanisms or the most scrutinizing eyes.

To sum up, some commonly used methods for recycling and recovering non-ferrous metals are sensor-based methods which rely on the use of sensors to detect and sort specific metals, precipitation methods mainly used for recovery from aqueous solutions and wastewater treatment, and electrowinning that has broad application across many different industries.

PLASTIC RECYCLING

Plastic is among the most popular and important materials used in the modern world. However, its popularity is part of the huge problem and reason why plastics should be recycled. Instead of throwing them away polluting the land and our water bodies, we can optimize the lifespan of plastics by recycling and reusing them.

Plastic recycling refers to the process of recovering waste or scrap plastic and reprocessing it into useful product. Due to the fact that plastic is non-biodegradable, it is essential that it is recycled as part of the global efforts to reducing plastic and other solid waste in the environment.

Process of Plastic Recycling

Collection

Plastics are available in a number of forms for example plastic containers, jars, bottles, plastic bags, packaging plastic, big industrial plastics just to mention but a few. Due to their nature and availability, there are plastic collection centres and some business people have ventured into plastic collecting business as a source of income. Tons and tons of scrap plastic are collected and sent to a collecting yard where they are then packed and transported to plastic processing plants. Unfortunately, not all countries have the capacity to recycle plastic. Very few developing countries can actually recycle plastic. This means that, plastic waste is still a major problem to some countries in the world.

Sorting

The actual plastic recycling process starts with sorting of the different plastic items by their resin content and color. This process is also done to ensure all contaminates are eliminated. There are specially designed machines that help in sorting of the plastics according to their resin content. Then the recycling mill sorts the scrap plastic by symbols at the bottom of the plastics.

Shredding

After sorting the plastics, the next step is to cut the plastics into tiny chunks or pieces. The plastic bottles and containers are then ground and cut into tiny pieces or flakes. The heavier and lighter plastic flakes are separated using a specially designed machine. The separation process helps in ensuring that the different plastics are not put together or mixed up in the final product. Remember that different plastics are used to make different items.

Cleaning

After a complete separation, the flakes or chunks are then washed with detergents to remove the remaining contamination. Once the cleaning process is complete, the clean flakes are passed through specialized equipment that further separates the plastic resin types. The plastic flakes are then subjected to moderate heat to dry.

Melting

The dry flakes are melted down. They can be melted down and molded into a new shape or they are melted down and processed into granules. The melting process is done under regulated temperatures. There is specialized equipment designed to melt down plastic without destroying them.

Making of Pellets

After the melting process, the plastic pieces are then compressed into tiny pellets known as nurdles. In this state, the plastic pellets are ready for reuse or be redesigned into new plastic products. It is important to point out that recycled plastic is hardly used to make identical plastic item or its previous form. It is in this pellet form that plastics are transported to plastic manufacturing companies to be redesigned and be used in making other useful plastic products.

What are the Common Recycled Plastics?

There are numerous and common types of recycled plastics as can be seen below:

Polyethylene Terephthalate

This type of recycled plastic is tough, has excellent clarity, is strong and has barrier to moisture and gas. It is used in the manufacture of water, soft drinks, peanut butter and salad dressing bottles and jars.

High Density Polyethylene

This recycled plastic is known for its excellent stiffness, resistance to moisture, strength, versatility, toughness and reduced permeability to gas. It is used in the manufacture of water, juice and milk bottles. It is also used to make retail and trash bags for households and business people.

Polyvinyl Chloride

Abbreviated as PVC, polyvinyl chloride has a number of applications. It is versatile, can be bended easily, it is tough and strong. This recycled plastic is commonly used in the manufacture of juice bottles, PVC piping and cling films.

Low Density Polyethylene

This is the most common type of recycled plastic. It has exceptional ease of processing; it is strong, flexible, tough, and resistant to moisture and it's easy to seal. This plastic is usually used in making frozen food bags, flexible container lids, freezable bottles just to mention but a few.

Advantages of Recycling Plastics

Plastics should be recycled because of a number of reasons as can be seen below:

Provision of a Sustainable Source of Raw Materials

Recycling plastics provides a sustainable source of raw materials to the manufacturing industry. Once the plastics are recycled, they are sent to manufacturing industries to be redesigned and converted into new shapes and used in different appliances.

Reduces Environmental Problems

Since plastics are non-biodegradable, they pose a high risk to the people and the environment as a whole. They can block sewer lines, drainages and other waterways leading to blockages and unwanted pileups. When plastics are eliminated through recycling, the environment looks clean and inhabitable.

Reduces Landfill Problems

Recycling plastics minimizes the amount of plastic being taken to the ever diminishing landfill sites. Most countries have designated areas specifically meant for burying plastics. When they are recycled, these sites will receive little plastic garbage. The remaining areas can be used for other purposes instead of dumping plastics that do not rot. These areas can be used for agriculture or for human settlement. It should be understood that human population is growing each day and land is becoming a problem. Instead of misusing the land for garbage disposal it can be used for settlement and other important economic activities.

Consumes Less Energy

Recycling of materials including plastics requires less energy as compared to making the plastic from scratch. This saves energy and that energy can be diverted to other important things in the economy. It is therefore important to encourage plastic recycling in the manufacturing industry as it will save the economy billions of money. The process of manufacturing plastic using natural raw materials is expensive and time consuming compared to the recycling process.

Encourages a Sustainable Lifestyle among People

Individuals who have ventured into plastic collection and recycling business will experience improved lifestyles as they will get their daily income from the business. This will in the long run improve the economy and boost the living standards of the people. So do not just sit there doing nothing, embrace plastic recycling activities and improve your economic standards.

Any sort of effort aimed at saving the environment is very important and matters a lot. Since its inception during the environmental revolution in the late 1960s, plastic recycling is one of the most encouraged solid waste management programs in the world. Prior to the push to use of plastic containers by manufacturers, products were packaged in glass, metal and paper. Therefore, in order to keep our environment clean, reduce landfills, provide a sustainable supply of plastics to manufacturers, it is important to recycle plastics.

PET Bottle Recycling

Bottles made of polyethylene terephthalate (PET, sometimes PETE) can be used to make lower grade products, such as carpets. To make a food grade plastic, the bottles need to be hydrolysed down to monomers, which are purified and then re-polymerised to make new PET. In many countries, PET plastics are coded with the resin identification code number "1" inside the universal recycling symbol, usually located on the bottom of the container.

Usage of PET

PET is used as a raw material for making packaging materials such as bottles and containers for packaging a wide range of food products and other consumer goods. Examples include soft drinks, alcoholic beverages, detergents, cosmetics, pharmaceutical products and edible oils. PET is one of the most common consumer plastics used. Polyethylene terephthalate can also be used as the main material in making water-resistant paper.

Process

"1-PETE" resin identification code.

Post-consumer Waste

The empty PET packaging is discarded by the consumer, after use and becomes PET waste. In the recycling industry, this is referred to as "post-consumer PET." Many local governments and waste collection agencies have started to collect post-consumer PET separately from other household waste. Besides that there is container deposit legislation in some countries which also applies to PET bottles.

It is debatable whether exporting circulating resources that damages the domestic recycling industry is acceptable or not. In Japan, overseas market pressure led to a significant cost reduction in the domestic market. The cost of the plastics other than PET bottles remained high.

Reverse vending machine for empty beverage cans and PET bottles in an Aldi supermarket.

Recycling bins usually include one for glass and plastic bottles, such as Urban Environmental Management and Technology.

Sorting

When the PET bottles are returned to an authorized redemption center, or to the original seller in some jurisdictions, the deposit is partly or fully refunded to the redeemer. In both cases the collected post-consumer PET is taken to recycling centres known as materials recovery facilities (MRF) where it is sorted and separated from other materials such as metal, objects made out of other rigid plastics such as PVC, HDPE, polypropylene, flexible plastics such as those used for bags (generally low density polyethylene), drink cartons, glass, and anything else which is not made out of PET.

Post-consumer PET is often sorted into different colour fractions: Transparent or uncoloured PET, blue and green coloured PET, and the remainder into a mixed colours fraction. The emergence of new colours (such as amber for plastic beer bottles) further complicates the sorting process for the recycling industry.

PET bottles are separated from other plastics in a materials recovery facility.

Bales of crushed PET bottles sorted according to color: blue, transparent, and green.

Bales of crushed PET bottles.

Processing for Sale

The sorted post-consumer PET waste is crushed, pressed into bales and offered for sale to recycling companies. Colourless/light blue post-consumer PET attracts higher sales prices than the darker blue and green fractions. The mixed color fraction is the least valuable.

Further Treatment

The further treatment process includes crushing, washing, separating and drying. Recycling companies further treat the post-consumer PET by shredding the material into small fragments. These fragments still contain residues of the original content, shredded paper labels and plastic caps. These are removed by plastic granulation, resulting in pure PET fragments, or "PET flakes". PET flakes are used as the raw material for a range of products that would otherwise be made of polyester. Examples include polyester fibres (a base material for the production of clothing, pillows, carpets, etc.), polyester sheets, strapping, or back into PET bottles.

Melt Filtration

Melt filtration is typically used to remove contaminants from polymer melts during the extrusion process. There is a mechanical separation of the contaminants within a machine called a 'screen changer'. A typical system will consist of a steel housing with the filtration medium contained in moveable pistons or slide plates that enable the processor to remove the screens from the extruder flow without stopping production. The contaminants are usually collected on woven wire screens which are supported on a stainless steel plate called a 'breaker plate'—a strong circular piece of steel drilled with large holes to allow the flow of the polymer melt. For the recycling of polyester it is typical to integrate a screen changer into the extrusion line. This can be in a pelletizing, sheet extrusion or strapping tape extrusion line.

Drying Polyester

PET polymer is very sensitive to hydrolytic degradation, resulting in severe reduction in its molecular weight, thereby adversely affecting its subsequent melt processability. Therefore, it is essential to dry the PET flakes or granules to a very low moisture level prior to melt extrusion. PET must be dried to <100 parts per million (ppm) moisture and maintained at this moisture level to minimize hydrolysis during melt processing.

Dehumidifying Drying: These types of dryers circulate hot and de-humidified dry air onto the resin, suck the air back, dry it and then pump again in a closed loop operation. This process reduces moisture level in the PET down to 50ppm or lower. The efficiency of moisture removal depends on the air dew point. If the air dew point is not good, then some moisture remains in the chips and cause IV loss during processing.

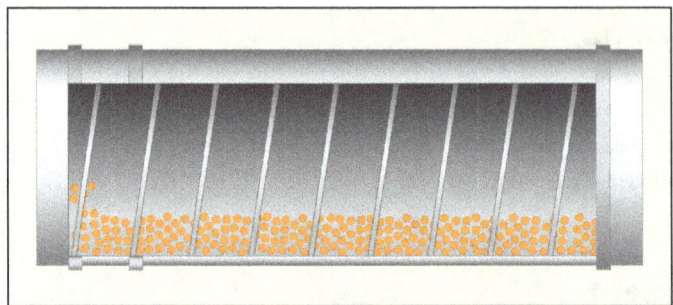

IRD drying drum used for the drying of
Polyester pellets and polyester bottle flakes.

Infrared Drying polyester pellets and flakes: A new type of dryer has been introduced in recent years, using Infrared drying (IRD). Due to the high rate of energy transfer with IR heating in combination with the specific wavelength used, the energy costs involved with these systems can be greatly reduced, along with the size. Polyester can be dried and amorphous flake crystallized and dried within only about 15 minutes down to a moisture level of approx. 300ppm in one step, and down to <50 ppm using a buffer hopper to complete the drying in typically under 1 hour.

Global Statistics

Dog raincoat made from 100% recycled PET fabric.

Worldwide, approximately 7.5 million tons of PET were collected in 2011. This gave 5.9 million tons of flake. In 2009 3.4 million tons were used to produce fibre, 500,000 tons to produce bottles, 500,000 tons to produce APET sheet for thermoforming, 200,000 tons to produce strapping tape and 100,000 tons for miscellaneous applications.

Petcore, the European trade association that fosters the collection and recycling of PET, reported that in Europe alone, 1.6 million tonnes of PET bottles were collected in 2011 - more than 51% of all bottles. After exported bales were taken into account, 1.12 million tons of PET flake were produced. 440,000 tons were used to produce fibres, 283,000 tons to produce more bottles, 278,000 tons to produce APET sheets, 102,000 tons for strapping tape and 18,000 tons for miscellaneous applications.

In 2008 the amount of post-consumer PET bottles collected for recycling and sold in the United States was approx. 1.45 billion pounds. In 2012, 81% of the PET bottles sold in Switzerland were recycled.

In 2018, 90% of the PET bottles sold in Finland were recycled. The high rate of recycling is mostly result of the deposit system in use. The law demands a tax of 0,51 €/l for bottles and cans that are not part of a refund system. Thus encouraged by the law, products are included to have a 10¢ to 40¢ deposit that is paid to the recycler of the can or bottle.

Increasing energy prices may increase the volume of recycling PET bottles. In Europe, the EU Waste Framework Directive mandates that by 2020 there should be 50% recycling or reuse of plastics from household streams.

In the United States the recycling rate for PET packaging was 31.2% in 2013, according to a report from The National Association for PET Container Resources (NAPCOR) and The Association of Postconsumer Plastic Recyclers (APR). A total of 1,798 million pounds was collected and 475 million pounds of recycled PET used out of a total of 5,764 million pounds of PET bottles.

PET Bottles Recycle-rate Globally

Japan	US	Europe	India
72%	29%	48%	90%

Re-use of PET Bottles

PET bottles are also recycled as-is (re-used) for various purposes, including for use in school projects, and for use in solar water disinfection in developing nations, in which empty PET bottles are filled with water and left in the sun to allow disinfection by ultraviolet radiation. PET is useful for this purpose because many other materials (including window glass) that are transparent to visible light are opaque to ultraviolet radiation.

A novel use is as a building material in third-world countries. According to online sources, the bottles, in a labor-intensive process, are filled with sand, then stacked and either mudded or cemented together to form a wall. Some of the bottles can be filled instead with air or water, to admit light into the structure.

Recycling of Thermosetting Plastics

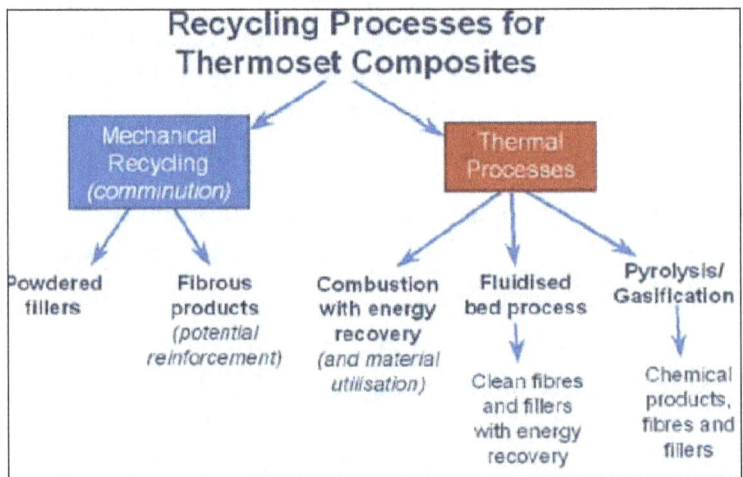

Thermoset composite materials are used in an increasingly wide range of applications and about one million tonnes are manufactured each year in Europe. Despite this success, recycling is a more difficult issue and the perception that composites are not recyclable is seen as a key barrier to their development in some markets.

Waste management is now a high priority, particularly in the European Union, and directives are restricting traditional disposal routes, such as landfill, and requiring recycling of other waste streams. For instance, the End- of-Life Vehicle Directive requires that by 2015 all vehicles disposed of must be 85% recyclable. Consequently, there is a need for recycling routes for composites to be established.

There are particular problems in trying to recycle composites. Thermosetting polymers are cross-linked and cannot be remoulded, unlike thermoplastics. Composites are by their very nature mixtures of different materials: polymer, fibre reinforcement and often fillers. There are few standard formulations and so compositions vary and composites are often manufactured integrally bonded with other materials such as foam cores or metal inserts.

Research and development of recycling techniques has been ongoing for some years and there are two fundamental categories of process for recycling thermoset composites, as shown in figure.

Mechanical Recycling

Mechanical recycling techniques involve grinding techniques to reduce the size of scrap material into powder or fibrous recyclates that can be used as raw materials. The technique is to use a primary crushing process to reduce the scrap components into manageable sized pieces. Then, a hammer mill or other high-speed mill is used to grind the material to a finer product, with particles ranging from fibrous strands, up to 10mm in length, down to fine powders of less than 50 microns in size. All the constituents of the original composite are reduced in size and appear in the recyclates, which are thus a mixture of polymer, fibre and filler.

The recyclates can be used at limited substitution levels in the manufacture of new, short-fibre composite moulding compounds such as sheet moulding compound (SMC), mainly as partial substitutes for the filler. Some coarser grades of recyclate containing more fibre can be used as partial substitutes for reinforcement. But in all cases, the mechanical properties deteriorate at high recyclate substitution levels. Many other uses for recyclate have been investigated. For example, a novel twin-screw extrusion process has been developed in which thermoset recyclate can be compounded with thermoplastics. And in Sweden, IFP SICOMP have developed a glass fibre base reinforcement (Recycore) where the core contains recyclate with a high permeability to allow resin flow during impregnation. Coarse grades of recyclate have also been used as reinforcement in asphalt and in the manufacture of plastic lumber from thermoplastics, where the recyclate can be used as an alternative to wood fibre.

Thermal Recycling

Thermal recycling techniques involve the use of heat to break down the composite. Thermosetting polymers have a calorific value similar to good quality coal and trials have shown the composites can successfully be burned for energy recovery, for example by mixing with municipal waste in an incinerator. Value can be recovered from the incombustible materials if the scrap is burned in a cement kiln, where glass fibres and mineral fillers can be used as raw materials for cement. The fibre reinforcement has potentially the most recoverable value in a composite and research has developed a fluidised bed process to recover high-grade fibre, shown in figure. In this process, pieces of scrap composite are fed into a bed of sand fluidised with hot air operating temperatures between typically 450 °C and 550 °C. At these temperatures, the thermosetting polymer volatilises and, once the polymer has been removed, the fibres and any mineral fillers are released and carried away in the gas stream. They are then separated out whilst the gases are fed to a high temperature combustion chamber for full oxidation and energy recovery. A high-quality fibre recyclate is produced which is clean, has good mechanical properties and has potential for reuse in applications requiring disperse short fibre such as short-fibre moulding compounds – both thermoset and thermoplastic – and non-woven fabrics. The advantage of the process is that it is tolerant of mixed and contaminated materials.

For instance, high-grade glass fibres were recycled from a composite component comprising two painted GRP skins with a foam core. Even the metal inserts did not have to be removed and were recovered from the fluidised bed. Research is currently focusing on the recycling of more valuable carbon fibre composites and a new project has recently commenced in which the recycling of carbon fibre vessels used for the storage of hydrogen fuel in vehicles is being investigated and the process is being developed to recover chemical products rather than energy from the polymer. Pyrolysis has also been investigated for recycling thermoset composites. In this process, the scrap composite is heated in the absence of air. The polymer decomposes to form lower molecular weight organic materials (liquids and gases) and a solid char product is also produced that is mixed with the recovered fibres. In a recent investigation potentially valuable chemicals were among the organic products formed, but further refining would be needed to separate them.

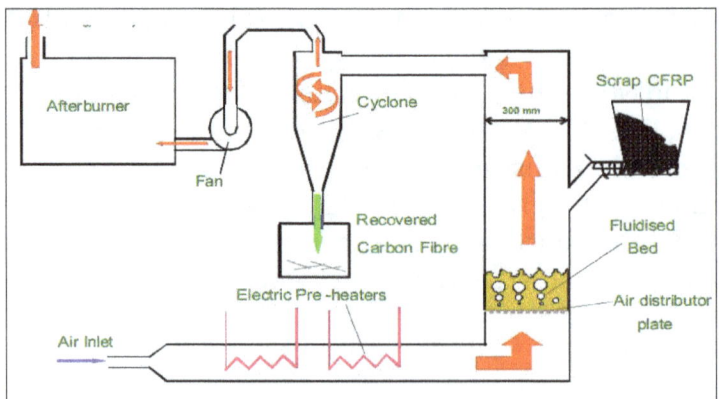

ALUMINIUM RECYCLING

Aluminium recycling is the process by which scrap aluminium can be reused in products after its initial production. The process involves simply re-melting the metal, which is far less expensive and energy-intensive than creating new aluminium through the electrolysis of aluminium oxide (Al_2O_3), which must first be mined from bauxite ore and then refined using the Bayer process. Recycling scrap aluminium requires only 5% of the energy used to make new aluminium from the raw ore. For this reason, approximately 36% of all aluminium produced in the United States comes from old recycled scrap. Used beverage containers are the largest component of processed aluminum scrap, and most of it is manufactured back into aluminium cans.

An aluminium recycling symbol.

The European Committee for Standardization logo for aluminium recycling.

A common practice since the early 1900s and extensively capitalized during World War II, aluminium recycling is not new. It was, however, a low-profile activity until the late 1960s, when the exploding popularity of aluminium beverage cans finally placed recycling into the public consciousness.

Model promoting aluminium recycling at Douglas Aircraft Company.

Sources for recycled aluminium include aircraft, automobiles, bicycles, boats, computers, cookware, gutters, siding, wire, and many other products that need a strong lightweight material, or a material with high thermal conductivity. As recycling does not transmute the element, aluminium can be recycled indefinitely and still be used to produce any product for which new aluminium could have been used.

Hydraulic press and baled cans prepared for transport.

Advantages

Aluminium is an infinitely recyclable material, and it takes up to 95 percent less energy to recycle it than to produce primary aluminum, which also limits emissions, including greenhouse gases. Today, about 75 percent of all aluminum produced in history, nearly a billion tons, is still in use.

The recycling of aluminium generally produces significant cost savings over the production of new aluminium, even when the cost of collection, separation and recycling are taken into account. Over the long term, even larger national savings are made when the reduction in the capital costs associated with landfills, mines, and international shipping of raw aluminium are considered.

Energy Savings

Recycling aluminium uses about 5% of the energy required to create aluminium from bauxite; the amount of energy required to convert aluminium oxide into aluminium can be vividly seen when the process is reversed during the combustion of thermite or ammonium perchlorate composite propellant.

Aluminium die extrusion is a specific way of getting reusable material from aluminium scraps but does not require a large energy output of a melting process. In 2003, half of the products manufactured with aluminium were sourced from recycled aluminium material.

Environmental Savings

The benefit with respect to emissions of carbon dioxide depends on the type of energy used. Electrolysis can be done using electricity from non-fossil-fuel sources, such as nuclear, geothermal, hydroelectric, or solar. Aluminium production is attracted to sources of cheap electricity. Canada, Brazil, Norway, and Venezuela have 61 to 99% hydroelectric power and are major aluminium producers. The use of recycled aluminium also decreases the need for mining bauxite.

The vast amount of aluminium used means that even small percentage losses are large expenses, so the flow of material is well monitored and accounted for financial reasons. Efficient production and recycling benefits the environment as well.

Process for Beverage Cans

Aluminium beverage cans are usually recycled by the following method:

- Cans are first divided from municipal waste, usually through an eddy current separator, and cut into small, equally sized pieces to lessen the volume and make it easier for the machines that separate them.

- Pieces are cleaned chemically/mechanically and blocked to minimize oxidation losses when melted. (The surface of aluminium readily oxidizes back into aluminium oxide when exposed to oxygen).

- Blocks are loaded into the furnace and heated to 750 °C ± 100 °C to produce molten aluminium.

- Dross is removed, and the dissolved hydrogen is degassed. (Molten aluminium readily disassociates hydrogen from water vapor and hydrocarbon contaminants). This is typically done with chlorine and nitrogen gas. Hexachloroethane tablets are normally used as the source for chlorine. Ammonium perchlorate can also be used, as it decomposes mainly into chlorine, nitrogen, and oxygen when heated.

- Samples are taken for spectroscopic analysis. Depending on the final product desired, high-purity aluminium, copper, zinc, manganese, silicon, and/or magnesium is added to alter the molten composition to the proper alloy specification. The top-five aluminium alloys produced are 6061, 7075, 1100, 6063, and 2024.

- The furnace is tapped, the molten aluminium poured out, and the process is repeated again for the next batch. Depending on the end product, it may be cast into ingots, billets, or rods, formed into large slabs for rolling, atomized into powder, sent to an extruder, or transported in its molten state to manufacturing facilities for further processing.

Ingot Production using Reverberatory Furnaces

The scrap aluminium is separated into a range of categories such as irony aluminium (engine blocks etc.), clean aluminium (alloy wheels). Scraps are classified according to ISRI (Institute of Scrap Recycling Industries).

Depending on the specification of the required ingot casting, it will depend on the type of scrap used in the start melt. Generally, the scrap is charged to a reverberatory furnace (other methods appear to be either less economical and/or dangerous) and melted down to form a "bath". The molten metal is tested using spectroscopy on a sample taken from the melt to determine what refinements are needed to produce the final casts. After the refinements have been added, the melt may be tested several times to be able to fine-tune the batch to the specific standard.

Once the correct "recipe" of metal is available, the furnace is tapped and poured into ingot moulds, usually via a casting machine. The melt is then left to cool, stacked and sold on as cast silicon–aluminium ingot to various industries for re-use. Mainly, cast alloys like ADC_{12}, LM_2, $AlSi_{132}$, LM_{24} etc. are produced. These secondary alloys ingots are used in die cast companies. Tilting rotary furnaces are used for recycling of aluminium scrap, which give higher recovery compared to reverberatory furnaces (Skelner Furnace).

Recycling Rates

Brazil recycles 98.2% of its aluminium can production, equivalent to 14.7 billion beverage cans per year, ranking first in the world, more than Japan's 82.5% recovery rate. Brazil has topped the aluminium can recycling charts eight years in a row.

Secondary Aluminium Recycling

White dross, a residue from primary aluminium production and secondary recycling operations, usually classified as waste, still contains useful quantities of aluminium which can be extracted industrially. The process produces aluminium billets, together with a highly complex waste material. This waste is difficult to manage. It reacts with water, releasing a mixture of gases (including, among others, hydrogen, acetylene, and ammonia) which spontaneously ignites on contact with air; contact with damp air results in the release of copious quantities of ammonia gas. Despite these difficulties, however, the waste has found use as a filler in asphalt and concrete.

CONCRETE RECYCLING

When structures made of concrete are demolished or renovated, concrete recycling is an increasingly common method of utilizing the rubble. Concrete was once routinely trucked to landfills for disposal, but recycling has a number of benefits that have made it a more attractive option in this age of greater environmental awareness, more environmental laws, and the desire to keep construction costs down.

Concrete aggregate collected from demolition sites is put through a crushing machine. Crushing facilities accept only uncontaminated concrete, which must be free of trash, wood, paper and other such materials. Metals such as rebar are accepted, since they can be removed with magnets and other sorting devices and melted down for recycling elsewhere. The remaining aggregate chunks are sorted by size. Larger chunks may go through the crusher again. After crushing has taken place, other particulates are filtered out through a variety of methods including hand-picking and water flotation.

Concrete from a building being sent to a portable crusher.
This is the first step to recycling concrete.

Crushing at the actual construction site using portable crushers reduces construction costs and the pollution generated when compared with transporting material to and from a quarry. Large road-portable plants can crush concrete and asphalt rubble at 600 tons per hour or more. These systems normally consist of a rubble crusher, side discharge conveyor, screening plant, and a return conveyor from the screen to the crusher inlet for reprocessing oversize materials. Compact, self-contained mini-crushers are also available that can handle up to 150 tons per hour and fit into tighter areas. With the advent of crusher attachments - those connected to various construction equipment, such as excavators - the trend towards recycling on-site with smaller volumes of material is growing rapidly. These attachments encompass volumes of 100 tons/hour and less.

Uses of Recycled Concrete

Smaller pieces of concrete are used as gravel for new construction projects. Sub-base gravel is laid down as the lowest layer in a road, with fresh concrete or asphalt poured over it. The US Federal Highway Administration may use techniques such as these to build new highways from the materials of old highways. Crushed recycled concrete can also be used as the dry aggregate for brand new concrete if it is free of contaminants. Also, concrete pavements can be broken in place and used as a base layer for an asphalt pavement through a process called rubblization.

Larger pieces of crushed concrete can be used as riprap revetments, which are "a very effective and popular method of controlling streambank erosion." With proper quality control at the crushing facility, well graded and aesthetically pleasing materials can be provided as a substitute for landscaping stone or mulch.

Wire gabions (cages), can be filled with crushed concrete and stacked together to provide economical retaining walls. Stacked gabions are also used to build privacy screen walls (in lieu of fencing).

Use of Recycled Coarse Aggregate in Concrete

Recent statistics show that the increasing demand of construction aggregate could reach 48.3 billion metric tons by the year 2015 with the highest consumption being in Asia and Pacific. The high demand of concrete means more new building will be constructed after the demolition of old buildings, generating a large volume of C&D waste (construction waste & demolition waste) as a by product of economic growth. However, the most common way to dispose this waste is by dumping it in a landfill. Without proper maintenance, landfills can cause many environmental problems such as air pollution and water contamination. This, along with the shortage of resources caused by this growth in construction, has caused more and more countries to begin considering the importance of C&D waste recycling. In general, the reuse and recycle of construction waste is concentrated in the preparation of recycled aggregate for concrete. By adding a portion of recycled aggregate instead of natural aggregate coarse into the mixture, producing the recycled concrete, which can conserve energy and materials for concrete production.

Strength and Durability of the Recycled Aggregate

Some experiments showed that recycled aggregate doesn't have good durability like the natural coarse aggregate but the durability can be improved by mixing it with special materials such as fly ash to produce high strength and durable concrete.

Benefits

There are a variety of benefits in recycling concrete rather than dumping it or burying it in a landfill:

- Save landfill space.
- Conserve natural resources by reducing the need for gravel mining, water, coal, oil and gas.
- When used as the base material for roadways, reduces pollution from waste transport to landfills and dumps.
- Create employment opportunities.
- Drags down material and waste transport expenses.
- Recycling one ton of cement could save 1,360 gallons water, 900 kg of CO_2.

Disadvantage of using Recycled Concrete

There have been concerns about the recycling of painted concrete due to possible lead content. The Army Corps of Engineers' Construction Engineering Research Laboratory (CERL) and others have conducted studies to see if lead-based paint in crushed concrete actually poses a hazard. It was concluded that concrete with lead-based paint would be able to be used as clean fill without impervious cover but with some type of soil cover.

COTTON RECYCLING

Cotton bolls on the cotton plant ready for harvesting and processing into cotton yarn and fabric.

The cotton recycling symbol.

Cotton recycling prevents unneeded wastage and can be a more sustainable alternative to disposal. Recycled cotton can come from secondhand clothing or from textile waste or leftovers which are then spun into new yarns and fabrics. There are some notable limitations of recycled cotton, including separation of materials that are cotton/polyester mix. There may also be limits to durability in using recycled cotton.

Process

Cotton can be recycled from pre-consumer (post-industrial) and post-consumer cotton waste. Pre-consumer waste comes from any excess material produced during the production of yarn, fabrics and textile products, e.g. selvage from weaving and fabric remnants from factory cutting rooms. Post-consumer waste comes from discarded textile products, e.g. used apparel and home textiles. During the recycling process, the cotton waste is first sorted by type and color and then processed through stripping machines that break the yarns and fabric into smaller pieces before pulling them apart into fiber. The mix is carded several times in order to clean and mix the fibers before they are spun into new yarns.

The resulting staple fiber is shorter than the original fiber length, meaning it is more difficult to spin. Recycled cotton is therefore often blended with virgin cotton fibers to improve yarn strengths. Commonly, not more than 30% recycled cotton content is used in the finished yarn or fabric.

Because waste cotton is often already dyed, re-dyeing may not be necessary. Cotton is an extremely resource-intense crop in terms of water, pesticides and insecticides. This means that using recycled cotton can lead to significant savings of natural resources and reduce pollution from agriculture. Recycling one tonne of cotton can save 765 cubic metres (202,000 US gal) of water.

Uses

Recycled cotton is often combined with recycled plastic bottles to make clothing and textiles, creating sustainable, earth-conscious products. Recycled cotton can also be used in industrial settings as polishing and wiper cloths and can even be made into new, high-quality paper. When reduced to its fibrous state, cotton can be used for applications like seat stuffing or home and automotive insulation. It is also sold as recycled cotton yarn for consumers to create their own items. Additionally, cotton waste can be made into a stronger, more durable paper than traditional wood-pulp based paper, which may contain high concentration of acids. Cotton paper is often used for important documents and also for bank notes since it does not wear off as easily. Cotton waste can also be used to grow mushrooms (particularly the indoor cultivation of Volvariella volvacea otherwise known as Straw Mushrooms).

Even though recycling cotton cuts down on the harsh process of creating brand new cotton products, it is a natural fiber and is biodegradable, so any cotton fibers that cannot be recycled or used further can be composted and will not take up space in landfills.

COPPER RECYCLING

Copper has been in use by civilization for over 10,000 years and has been recycled since early times. Because it does not degrade during recycling, copper in use today could have been first fabricated into objects thousands of years ago. Copper is highly prized by scrap metal collectors and scrap metal recycling businesses. The nonferrous metal is the best conductor of electricity except for silver. That electrical and thermal conductivity, along with properties of high ductility and malleability make it one of the most demanded metals by industry, eclipsed only by iron and aluminum.

Environmental Importance of Copper Recycling

Copper is an essential trace element that is necessary for plant and animal health. Moderate excess exposure to copper is not associated with health risks. As with other metals, there are significant environmental benefits to the recycling of copper. These include solid waste diversion, reduced energy requirements for processing, and natural resource conservation. For example, the energy requirements of recycled copper are as much as 85 to 90% less than the processing of new copper from virgin ore. In terms of conservation, copper is a non-renewable resource, although only 12% of known reserves have been consumed. Known U.S. reserves of copper are thought to total 1.6 billion metric tons, with production concentrated in Arizona, Utah, New Mexico, Nevada, and Montana. About 99% of domestic production is generated from 20 mines.

An emerging environmental challenge for copper is its use in the ever-increasing production of electrical products that still experience low recycling rates. This trend is changing for the better, however, through electronics recycling initiatives.

Economic Importance of Copper Recycling

Ranking immediately behind Chile in copper production, the United States is largely self-sufficient in copper supply. The U.S. produces roughly 8% of the world's copper generation. Almost half of U.S. copper output comes from recycled material, however. In 2010, U.S. recyclers processed 1.8 million metric tons of copper for domestic use and export, second only to aluminum among nonferrous metals, which saw 4.6 million tonnes recycled.

Slightly over one-half of recycled copper scrap is new scrap recovery including chips and machine turnings, with the rest being old post-consumer scraps such as electrical cable, old radiators, and plumbing tube.

Where to find Copper for Recycling?

For the scrap metal collector, an important source of scrap is an electrical cable, copper flashing, old radiators, and plumbing work. Copper from buildings is crucial, and content is estimated below:

- In houses (estimates for a 2,100 square foot residence):
  - 195 pounds: Building wire,
  - 151 pounds: Plumbing tube, fittings, valves,
  - 24 pounds: Plumbers' brass goods,
  - 47 pounds: Built-in appliances,
  - 12 pounds: Builders hardware,
  - 10 pounds: Other wire and tube.
- In an apartment of 1,000 square feet:
  - 125 pounds: Building wire,
  - 82 pounds: Plumbing tube, fittings, valves,
  - 20 pounds: Plumbers' brass goods,

- ◦ 38 pounds: Built-in appliances,

- ◦ 6 pounds: Builders hardware,

- ◦ 7 pounds: Other wire and tube.

- • In residential appliances:

- ◦ 52 pounds: Unitary air conditioner,

- ◦ 48 pounds: Unitary heat pump,

- ◦ 5.0 pounds: Dishwasher,

- ◦ 4.8 pounds: Refrigerator/freezer,

- ◦ 4.4 pounds: Clothes washer,

- ◦ 2.7 pounds: Dehumidifier,

- ◦ 2.3 pounds: Disposer,

- ◦ 2.0 pounds: Clothes dryer,

- ◦ 1.3 pounds: Range.

Copper Purity and Recycling

The Copper Development Association notes that were copper is combined with other materials such as tin or solder, it can be more economical to utilize it in applications where tin or lead is required, such as gunmetals or bronzes, rather than to remove these metals through refining. A scrap of this type commands a lower price than uncontaminated copper. Where contamination has extended "beyond acceptable limits," to quote the Copper Development Association, re-refining is necessary to recover pure copper.

Steps in Scrap Copper Recycling

Recycling is a very good method to help the environment and save energy. Almost anything can be recycled including metal. According to Environmental Protection Agency, metal accounts for 34.6 percent of all solid waste recycled, just second to Paper. As we all know, copper , as the most common metal, is in urgent needs all over the world. For recent years, copper recycling industry has enjoyed more popularity as the increasing concept of environmental protection. The process for recycling copper involves several key steps, and understanding the process may encourage people to maximize their recycling efforts.

Prosperous Copper Recycling Industry

Unlike recycled plastic or glass, recycled copper have wide and indefinite applicaton. Copper can be recycled more than once as the attribute and function will not change with the continuous processing. These natures makes recycling copper very advantageous and copper recycling industry very profitable and prosperous.

Key Steps for Copper Recycling

Step 1: Collection

Copper collecting is the first step for recycling. You can collect copper wire , copper cable and other tinny spare parts at home. For large items containing copper, such as old appliances or scrap vehicles, you can contact your local recycling center to make arrangements to have them collected.

Step 2: Sorting and Stripping/Granulator

The copper parts can be sorted according to many ways such as the processing methods. According to world market analysis, the main machines for recycling copper in physical way are copper stripping machine and copper granulator. These two machines have their own unique features and advantages. You can sort the copper items according to the type of wire or cable. If the copper items are single wire with less knots, you can just choose the wire stripping machine. If the copper items are complicated copper cables with winding knots and multiple stands, then they can be processed by the copper granulator.

Step 3: Shredding and Melting

If the copper items processed by copper granulator, the pure copper can be collected and packed to be further processed. While copper items stripped by the wire stripping machine needs to be shredded before packing. Because the entire copper wire is not so convenient to be transported to the melting device. When the copper reach the smelting facilities, the bales are fed into a furnace where they are heated until they become molten copper. The molten metal is then poured into casters or molds to form items needed.

Step 4: Cooling and Fabrication

Once the metal ingots have cooled and hardened, they are put into a machine that rolls them into smooth sheets, which form the basis for new metal materials. The recycled copper is without losing the elements and characteristics and so they can be reused as good as the original copper. The malleability of copper attributes and the strength remain the same and you can get the same objects made from the recycled copper to bring in prosperity in the country.

GLASS RECYCLING

Bottles in different colors.

Mixed color glass cullet.

Glass recycling is the processing of waste glass into usable products. Glass that is crushed and ready to be remelted is called cullet. There are two types of cullet: internal and external. Internal cullet is composed of defective products detected and rejected by a quality control process during the industrial process of glass manufacturing, transition phases of product changes (such as thickness and colour changes) and production offcuts. External cullet is waste glass that has been collected or reprocessed with the purpose of recycling. External cullet (which can be pre- or post-consumer) is classified as waste. The word "cullet", when used in the context of end-of-waste, will always refer to external cullet.

To be recycled, glass waste needs to be purified and cleaned of contamination. Then, depending on the end use and local processing capabilities, it might also have to be separated into different colors. Many recyclers collect different colors of glass separately since glass retains its color after recycling. The most common colours used for consumer containers are clear (flint) glass, green glass, and brown (amber) glass. Glass is ideal for recycling since none of the material is degraded by normal use.

Many collection points have separate bins for clear (flint), green and brown (amber). Glass re-processors intending to make new glass containers require separation by color, because glass tends to retain its color after recycling. If the recycled glass is not going to be made into more glass, or if the glass re-processor uses newer optical sorting equipment, separation by color at the collection point may not be required. Heat-resistant glass, such as Pyrex or borosilicate glass, must not be part of the glass recycling stream, because even a small piece of such material will alter the viscosity of the fluid in the furnace at remelt.

Processing of External Cullet

To be able to use external cullet in production, any contaminants should be removed as much as possible. Typical contaminations are:

- Organics: Paper, plastics, caps, rings, PVB foils for flat glass.

- Inorganics: Stones, ceramics, porcelains.

- Metals: Ferrous and non-ferrous metals.

- Heat resistant and lead glass.

Manpower or machinery can be used in different stages of purification. Since they melt at higher temperatures than glass, separation of inorganics, the removal of heat resistant glass and lead glass is critical. In the modern recycling facilities, dryer systems and optical sorting machines are used. The input material should be sized and cleaned for the highest efficiency in automatic sorting. More than one free fall or conveyor belt sorter can be used, depending on the requirements of the process. Different colors can be sorted by optical sorting machines.

Recycling into Glass Containers

A variant of the "Tidyman" symbol,
intended to encourage people to recycle glass.

Glass bottles and jars are infinitely recyclable. The use of recycled glass in manufacturing conserves raw materials and reduces energy consumption. Because the chemical energy required to melt the raw materials has already been expended, the use of cullet can significantly reduce energy consumption compared with manufacturing new glass from silica (SiO_2), soda ash ($Na2CO_3$), and lime ($CaCO_3$). Soda lime glass from virgin raw materials theoretically requires approximately 2.671 GJ/tonne compared to 1.886 GJ/tonne to melt 100% glass cullet. As a general rule, every 10% increase in cullet usage results in an energy savings of 2–3% in the melting process, with a theoretical maximum potential of 30% energy saving. Every metric ton (1,000 kg) of waste glass recycled into new items saves 315 kilograms (694 lb) of carbon dioxide from being released into the atmosphere during the manufacture of new glass.

Recycling into other Products

The use of the recycled glass as aggregate in concrete has become popular, with large-scale research on that application being carried out at Columbia University in New York. Recycled glass greatly enhances the aesthetic appeal of the concrete. Recent research has shown that concrete made with recycled glass aggregates have better long-term strength and better thermal insulation, due to the thermal properties of the glass aggregates. Glass which is not recycled, but crushed, reduces the volume of waste sent to landfill. Waste glass may also be kept out of landfill by using it for roadbed aggregate or landfill cover.

Glass aggregate, a mix of colors crushed to a small size, is substituted for pea gravel or crushed rock in many construction and utility projects, saving municipalities, such as the City of Tumwater, Washington Public Works, thousands of dollars (depending on the size of the project). Glass aggregate is not sharp to handle. In many cases, the state Department of Transportation has specifications for use, size and percentage of quantity for use. Common applications are as pipe bedding—placed around sewer, storm water or drinking water pipes, to transfer weight from the surface and protect the pipe. Another common use is as fill to bring the level of a concrete floor even with a foundation.

Other uses for recycled glass include:

- Fiberglass insulation products,
- Ceramic sanitary ware production,
- As a flux in brick manufacture,
- Astroturf,
- Agriculture and landscape applications, such as top dressing, root zone material or golf bunker sand,
- Recycled glass countertops,
- As water filtration media,
- Abrasives.

Mixed waste streams may be collected from materials recovery facilities or mechanical biological treatment systems. Some facilities can sort mixed waste streams into different colours using electro-optical sorting units.

Wordwide Process of Recycling

In 2019, many Australian cities after decades of poor planning and minimum investment are winding back their glass recycling programmes in favour of plastic usage.

For many years, there was only one state in Australia with a return deposit scheme on glass containers. Other states had unsuccessfully tried to lobby for glass deposit schemes. More recently this situation has changed dramatically, with the original scheme in South Australia now joined by legislated container deposit schemes in New South Wales, Queensland, Australian Capital Territory, and the Northern Territory, with schemes planned in Western Australia and Tasmania. Victoria is the only state not yet planning to introduce such a scheme.

In 2004, Germany recycled 2.116 million tons of glass. Reusable glass or plastic (PET) bottles are available for many drinks, especially beer and carbonated water as well as soft drinks (*Mehrwegflaschen*). The deposit per bottle (*Pfand*) is €0.08-€0.15, compared to €0.25 for recyclable but not reusable plastic bottles. There is no deposit for glass bottles which do not get refilled.

Vehicle emptying a glass recycling container in Vienna.

Glass collection points, known as *bottle banks* are very common near shopping centres, at civic amenity sites and in local neighborhoods in the United Kingdom. The first bottle bank was introduced by Stanley Race CBE, then president of the Glass Manufacturers' Federation and Ron England in Barnsley on 6 June 1977. Development work was done by the DoE at Warren Springs laboratory, Stevenage, (now AERA at Harwell) and Nazeing Glass Works, Broxbourne to prove if a usable glass product could be made from over 90% recycled glass. It was found necessary to use magnets to remove unwanted metal closures in the mixture.

Bottle banks commonly stand beside collection points for other recyclable waste like paper, metals and plastics. Local, municipal waste collectors usually have one central point for all types of waste in which large glass containers are located. There are now over 50,000 bottle banks in the United Kingdom, and 752,000 tons of glass are now recycled annually.

The waste recycling industry in the UK cannot consume all of the recycled container glass that will become available over the coming years, mainly due to the colour imbalance between that which is manufactured and that which is consumed. The UK imports much more green glass in the form of wine bottles than it uses, leading to a surplus amount for recycling.

The resulting surplus of green glass from imported bottles may be exported to producing countries, or used locally in the growing diversity of secondary end uses for recycled glass. As of 2006, Cory Environmental were shipping glass cullet from the UK to Portugal.

Rates of recycling and methods of waste collection vary substantially across the United States because laws are written on the state or local level and large municipalities often have their own unique systems. Many cities do curbside recycling, meaning they collect household recyclable waste on a weekly or bi-weekly basis that residents set out in special containers in front of their homes and transported to a materials recovery facility. This is typically single-stream recycling, which creates an impure product and partly explains why, as of 2019, the US has a recycling rate of around 33% versus 90% in some European countries.

Apartment dwellers usually use shared containers that may be collected by the city or by private recycling companies which can have their own recycling rules. In some cases, glass is specifically separated into its own container because broken glass is a hazard to the people who later manually sort the co-mingled recyclables. Sorted recyclables are later sold to companies to be used in the manufacture of new products.

In 1971, the state of Oregon passed a law requiring buyers of carbonated beverages (such as beer and soda) to pay five cents per container (increased to ten cents in April 2017) as a deposit which would be refunded to anyone who returned the container for recycling. This law has since been copied in nine other states including New York and California. The abbreviations of states with deposit laws are printed on all qualifying bottles and cans. In states with these container deposit laws, most supermarkets automate the deposit refund process by providing machines which will count containers as they are inserted and then print credit vouchers that can be redeemed at the store for the number of containers returned. Small glass bottles (mostly beer) are broken, one-by-one, inside these deposit refund machines as the bottles are inserted. A large, wheeled hopper (very roughly 1.5 m by 1.5 m by 0.5 m) inside the machine collects the broken glass until it can be emptied by an employee. Nationwide bottle refunds recover 80% of glass containers that require a deposit.

Major companies in the space include Strategic Materials, which purchases post-consumer glass for 47 facilities across the country. Strategic Materials has worked to correct misconceptions about glass recycling. Glass manufacturers such as Owens-Illinois ultimately include recycled glass in their product. The Glass Recycling Coalition is a group of companies and stakeholders working to improve glass recycling.

GYPSUM RECYCLING

Gypsum recycling is the process of turning gypsum waste into recycled gypsum, thereby generating a raw material that can replace virgin gypsum raw materials in the manufacturing of new products.

Gypsum waste primarily consists of waste from gypsum boards, which are wall or ceiling panels made of a gypsum core between paper lining. Such boards are also referred to as plasterboards, drywall, wallboards and gyprock. Gypsum waste in some countries also consists of gypsum blocks and plaster, among others.

Three main types of gypsum waste based on their origin can be distinguished:

- Gypsum waste from the manufacturing of gypsum products: This waste, which arises at the industrial gypsum production sites, consists of rejects and non-spec materials generated during the manufacturing of gypsum products. The recycling of this waste stream is usually part of the waste avoidance activity of the gypsum plants. The waste is referred to as gypsum manufacturing or production waste and the recycled gypsum obtained from the recycling of this is known as "production waste derived recycled gypsum".

- Gypsum waste from new construction: Gypsum waste from new construction activities is typically a clean waste, and primarily consists of off-cuts of plasterboard (drywall, wallboard or gyprock) when the boards have been cut to fit the dimensions of the wall or ceiling. The waste may constitute 15% of the gypsum materials used on the site. Such waste is generally referred to as new construction gypsum waste, and can be reduced by ordering boards "made-to-measure", but in most markets less than 10% of all orders are "made-to-measure".

- Gypsum waste from demolition and reconstruction: This waste arises when already installed plasterboards (drywalls, wallboards or gyprock boards), that usually have been installed many years ago, are taken out in connection with that the building is demolished or renovated. For this reason some refer to this waste as "old gypsum waste", whereas the trade usually refer to this waste as "demolition waste". Different from the two other types of gypsum waste described above, this type of gypsum waste from renovation, refurbishment and demolition works is more likely to present a certain degree of contamination, which can be in the form of nails, screws, wood, insulation, wall coverings etc. For this waste to be recyclable it is required that the equipment processing the waste is capable of separating such contamination from the gypsum to arrive at a pure recycled gypsum. New construction and demolition gypsum waste is both arising after the gypsum products have left the manufacturing sites, and together these two waste types are referred to as post consumer gypsum waste. The recycled gypsum obtained from this is known as post-consumer re-cycled gypsum.

Gypsum Recycling Process

Gypsum waste can be turned into recycled gypsum by processing the gypsum waste in such a way that the contaminants are removed and the paper facing of the plasterboard is separated from the gypsum core through mechanical processes including grinding and sieving in specialised equipment. Gypsum waste such as gypsum blocks and plaster do not require the removal of paper, as they are not made with paper from the beginning. It is typical for the gypsum recyclers to accept up to 3 per cent of contamination from other materials. The professional recyclers are capable of handling gypsum waste with nails and screws, wall coverings etc.

Need of Gypsum Recycling

Gypsum materials consist of calcium sulfate dihydrate ($CaSO_4 \cdot 2H_2O$). Sulfate-reducing bacteria convert sulfates to toxic hydrogen sulphide gas; they are killed by exposure to air, but the moist, airless, carbon-containing environment in a landfill is a good habitat for them. So gypsum put into landfill will decompose, releasing up to a quarter its weight in hydrogen sulfide. Moreover, methanogenic bacteria also thrive in such an environment, and convert the paper in the plasterboard to methane gas which is a potent greenhouse gas.

Recycling gypsum waste also reduces the need for the quarrying and production of virgin gypsum raw materials. Recycling one ton of the ordinary gypsum will save 1,000 pounds of black alkali, 1 ton of lactic acid and 500 kwh of energy. Recycling one metric ton of gypsum will save 28 kwh of energy and 4 pounds of aluminium.

Rationale for Choosing Closed Loop Recycling

Gypsum is fully and eternally recyclable and, as a consequence, gypsum waste is one of the few construction materials for which closed loop recycling is possible. Closed loop recycling of gypsum products involves the collection and processing of the gypsum waste, and the delivery of the obtained recycled gypsum to the manufacturer of gypsum products. It is therefore essential that the recycled gypsum achieves a pre-determined quality suitable for the manufacturing of new gypsum products. Presently there is no European or American standard pre-determining the recycled gypsum's quality and the criteria vary from plant to plant.

By choosing closed loop recycling the need for manufacturers to acquire virgin gypsum is reduced, contributing therefore to promote a sustainable manufacturing process. The most advanced plants, and most of these are found in the Nordic countries in Europe, have substituted up to 30 per cent of virgin gypsum raw materials with recycled gypsum.

PAPER RECYCLING

Waste paper collected for recycling in Italy.

Bin to collect paper for recycling in a German train station.

The recycling of paper is the process by which waste paper is turned into new paper products. It has a number of important benefits: It saves waste paper from occupying homes of people and producing methane as it breaks down. Because paper fibre contains carbon (originally absorbed by the tree from which it was produced), recycling keeps the carbon locked up for longer and out of the atmosphere. Around two thirds of all paper products in the US are now recovered and recycled, although it does not all become new paper. After repeated processing the fibres become too short for the production of new paper - this is why virgin fibre (from sustainably farmed trees) will be added to the pulp recipe.

There are three categories of paper that can be used as feedstocks for making *recycled paper*: Mill broke, pre-consumer waste, and post-consumer waste. *Mill broke* is paper trimmings and other paper scrap from the manufacture of paper, and is recycled in a paper mill. *Pre-consumer waste* is a material which left the paper mill but was discarded before it was ready for consumer use. *Post-consumer* waste is material discarded after consumer use, such as old corrugated containers (OCC), old magazines, and newspapers. Paper suitable for recycling is called "scrap paper", often used to produce moulded pulp packaging. The industrial process of removing printing ink from paper fibres of recycled paper to make deinked pulp is called deinking, an invention of the German jurist Justus Claproth.

Process

The process of waste paper recycling most often involves mixing used/old paper with water and chemicals to break it down. It is then chopped up and heated, which breaks it down further into strands of cellulose, a type of organic plant material; this resulting mixture is called pulp, or slurry. It is strained through screens, which remove any glue or plastic (especially from plastic-coated paper) that may still be in the mixture then cleaned, de-inked (ink is removed), bleached, and mixed with water. Then it can be made into new recycled paper. The share of ink in a wastepaper stock is up to about 2% of the total weight.

Rationale for Recycling

Industrialized paper making has an effect on the environment both upstream (where raw materials are acquired and processed) and downstream (waste-disposal impacts). Today, 40% of paper pulp is created from wood (in most modern mills only 9-16% of pulp is made from pulp logs; the rest comes from waste wood that was traditionally burnt). Paper production accounts for about 35% of felled trees, and represents 1.2% of the world's total economic output. Recycling one ton of newsprint saves about 1 ton of wood while recycling 1 ton of printing or copier paper saves slightly more than 2 tons of wood.

This is because kraft pulping requires twice as much wood since it removes lignin to produce higher quality fibres than mechanical pulping processes. Relating tons of paper recycled to the number of trees not cut is meaningless, since tree size varies tremendously and is the major factor in how much paper can be made from how many trees. In addition, trees raised specifically for pulp production account for 16% of world pulp production, old growth forests 9% and second- and third- and more generation forests account for the balance. Most pulp mill operators practice refor-estation to ensure a continuing supply of trees. The Programme for the Endorsement of Forest Certification (PEFC) and the Forest Stewardship Council (FSC) certify paper made from trees harvested according to guidelines meant to ensure good forestry practices.

Energy

Energy consumption is reduced by recycling, although there is debate concerning the actual energy savings realized. The Energy Information Administration claims a 40% reduction in energy when paper is recycled versus paper made with unrecycled pulp, while the Bureau of International Recycling (BIR) claims a 64% reduction. Some calculations show that recycling one ton of newspaper saves about 4,000 kWh (14 GJ) of electricity, although this may be too high. This is enough electricity to power a 3-bedroom European house for an entire year, or enough energy to heat and air-condition the average North American home for almost six months. Recycling paper to make pulp actually consumes more fossil fuels than making new pulp via the kraft process; these mills generate most of their energy from burning waste wood (bark, roots, sawmill waste) and byproduct lignin (black liquor). Pulp mills producing new mechanical pulp use large amounts of energy; a very rough estimate of the electrical energy needed is 10 gigajoules per tonne of pulp (2500 kW·h per short ton).

Landfill Use

About 35% of municipal solid waste (before recycling) in the United States by weight is paper and paper products. 42.4% of that is recycled.

Water and Air Pollution

The United States Environmental Protection Agency (EPA) has found that recycling causes 35% less water pollution and 74% less air pollution than making virgin paper. Pulp mills can be sources of both air and water pollution, especially if they are producing bleached pulp. Modern mills produce considerably less pollution than those of a few decades ago. Recycling paper provides an alternative fibre for papermaking. Recycled pulp can be bleached with the same chemicals used to bleach virgin pulp, but hydrogen peroxide and sodium hydrosulfite are the most common bleaching agents. Recycled pulp, or paper made from it, is known as PCF (process chlorine free) if no chlorine-containing compounds were used in the recycling process. However, recycling mills may have polluting by-products like sludge. De-inking at Cross Pointe's Miami, Ohio mill results in sludge weighing 22% of the weight of wastepaper recycled.

Recycling Facts and Figures

In the mid-19th century, there was an increased demand for books and writing material. Up to that time, paper manufacturers had used discarded linen rags for paper, but supply could not keep up with the increased demand. Books were bought at auctions for the purpose of recycling fiber content into new paper, at least in the United Kingdom, by the beginning of the 19th century.

Internationally, about half of all recovered paper comes from converting losses (pre-consumer recycling), such as shavings and unsold periodicals; approximately one third comes from household or post-consumer waste.

Some statistics on paper consumption:

- In 1996 it was estimated that 95% of business information is still stored on paper.

- Recycling 1 short ton (0.91 t) of paper saves 17 mature trees, 7 thousand US gallons (26 m³) of water, 3 cubic yards (2.3 m³) of landfill space, 2 barrels of oil (84 US gal or 320 l), and 4,100 kilowatt-hours (15 GJ) of electricity – enough energy to power the average American home for six months.

- Although paper is traditionally identified with reading and writing, communications has now been replaced by packaging as the single largest category of paper use at 41% of all paper used.

- 115 billion sheets of paper are used annually for personal computers. The average web user prints 16 pages daily.

- Most corrugated fiberboard boxes have over 25% recycled fibers. Some are 100% recycled fiber.

- In 1997, 299,044 metric tons of paper was produced (including paperboard).

- In the United States, the average consumption of paper per person in 1999 was approximately 354 kilograms. This would be the same consumption for 6 people in Asia or 30 people in Africa.

- In 2006-2007, Australia 5.5 million tons of paper and cardboard was used with 2.5 million tons of this recycled.

- Newspaper manufactured in Australia has 40% recycled content.

By Region

Paper recycling in Europe has a long history. The industry self-initiative European Recovered Paper Council(ERPC) was set up in 2000 to monitor progress towards meeting the paper recycling targets set out in the 2000 European Declaration on Paper Recycling. Since then, the commitments in the Declaration have been renewed every five years. In 2011, the ERPC committed itself to meeting and maintaining both a voluntary recycling rate target of 70% in the then E-27 plus Switzerland and Norway by 2015 as well as qualitative targets in areas such as waste prevention, ecodesign and research and development. In 2014 the paper recycling rate in Europe was 71.7%, as stated in the 2014 Monitoring Report.

Municipal collections of paper for recycling are in place. However, according to the *Yomiuri Shimbun*, in 2008, eight paper manufacturers in Japan have admitted to intentionally mislabeling recycled paper products, exaggerating the amount of recycled paper used.

Recycling has long been practiced in the United States. In 2012, paper and paperboard accounted for 68 million tons of municipal solid waste generated in the U.S., down from more than 87 million tons in 2000, according to the U.S. Environmental Protection Agency. While paper is the most commonly recycled material—64.6 percent was recovered in 2012—it is being used less overall than at the turn of the century. Paper accounts for more than a half of all recyclables collected in the US, by weight.

The history of paper recycling has several dates of importance:

- In 1690: The first paper mill to use recycled linen was established by the Rittenhouse family.

- In 1896: The first major recycling center was started by the Benedetto family in New York City, where they collected rags, newspaper, and trash with a pushcart.

- In 1993: The first year when more paper was recycled than was buried in landfills.

Today, over half of all paper used in the United States is collected and recycled. Paper products are still the largest component of municipal solid waste, making up more than 40% of the composition of landfills. In 2006, a record 53.4% of the paper used in the US (53.5 million tons) was recovered for recycling, up from a 1990 recovery rate of 33.5%. The US paper industry set a goal of recovering 55 percent of all paper used in the US by 2012. Paper products used by the packaging industry were responsible for about 77% of packaging materials recycled, with more than 24 million pounds recovered in 2005.

By 1998, some 9,000 curbside recycling programs and 12,000 recyclable drop-off centers existed nationwide. As of 1999, 480 materials recovery facilities had been established to process the collected materials. Recently, junk mail has become a larger part of the overall recycling stream, compared to newspapers or personal letters. However, the increase in junk mail is still smaller compared to the declining use of paper from those sources.

In 2008, the global financial crisis caused the price of old newspapers to drop in the U.S. from $130 to $40 per short ton ($140/t to $45/t) in October. In Mexico, recycled paper, rather than wood pulp, is the principal feedstock in papermills accounting for about 75% of raw materials. In 2018, South Africa recovered 1.285 million tonnes of recyclable paper products, putting the country's paper recovery rate at 71.7%. More than 90% of this recovered paper is used for the local beneficiation of new paper packaging and tissue.

Limitations and Impacts

Along with fibres, paper can contain a variety of inorganic and organic constituents, including up to 10,000 different chemicals, which can potentially contaminate the newly manufactured paper products. As an example, bisphenol A (a chemical commonly found in thermal paper) has been verified as a contaminant in a variety of paper products resulting from paper recycling. Furthermore, groups of chemicals as phthalates, phenols, mineral oils, polychlorinated biphenyls (PCBs) and toxic metals have all been identified in paper material. Although several measures might reduce the chemical load in paper recycling (e.g., improved decontamination, optimized collection of paper for recycling), even completely terminating the use of a particular chemical (phase-out) might still result in its circulation in the paper cycle for decades.

TIMBER RECYCLING

Timber recycling or wood recycling is the process of turning waste timber into usable products. Recycling timber is a practice that was popularized in the early 1990s as issues such as deforestation and climate change prompted both timber suppliers and consumers to turn to a more sustainable timber source. Recycling timber is the environmentally friendliest form of timber production and is very common in countries such as Australia and New Zealand where supplies of old wooden structures are plentiful. Timber can be chipped down into wood chips which can be used to power homes or power plants.

Demolishers pulling timber from an old wool store in Sydney, which will later be re-used for timber flooring.

Benefits

Example of recycled timber as a final product.

Recycling timber has become popular due to its image as an environmentally friendly product. Common belief among consumers is that by purchasing recycled wood, the demand for "green timber" will fall and ultimately benefit the environment. Greenpeace also view recycled timber as an environmentally friendly product, citing it as the most preferable timber source on their website. The arrival of recycled timber as a construction product has been important in both raising industry and consumer awareness towards deforestation and promoting timber mills to adopt more environmentally friendly practices. Recycling one ton of wood can save 18,000,000 btus of energy.

Drawbacks

Some hurdles facing the widespread adoption of recycled timber: sometimes the ends of wall studs need to be trimmed off to stop decay and cracking, thus resulting in a shorter piece of wood; this trimming may result in pieces of wood that do not meet building codes. Though the price may be less than for new wood, the process of selecting usable pieces of salvaged wood, pulling out nails, and refinishing for a new use can be laborious and time-consuming. Demolition must happen in such a way as to preserve as much of the timber as possible in a building, which means more time spent dismantling a building rather than just tearing it down quickly. The trade in recycled timber is not well-established everywhere, so a reliable supply of usable wood may be hard to come by for builders. There may be a stigma associated with using "used" or "cheap" wood that is perceived to be of not as high quality as "new" wood. Not all pieces of wood in a dismantled building will fit in a new building, and it may be cheaper and easier, from a design and labor perspective, to simply get new wood (ex: wood from a 6-foot (1.8 m) deck being used in a 7-foot (2.1 m) deck). Of course, none of these issues are insurmountable, and they are issues of convenience and logistics rather than structural integrity, but many builders find it easier and less time-consuming to simply get new wood in standard uniform sizes.

Recycling Timber

Recycled timber salvaged from a demolished building in Trat,
Thailand, is applied to make a new roof.

Recycled timber most commonly comes from old buildings, bridges and wharfs, where it is carefully stripped out and put aside by demolishers. At the same time any usable dimension stone is set aside for reuse. The demolishers then sell the salvaged timber to merchants who then re-mill the timber by manually scanning it with a metal detector, which allows the timber to be de-nailed and sawn to size. Once re-milled the timber is commonly sold to consumers in the form of timber flooring, beams and decking.

Examples:

Use of recycling timber is not new. As early as 1948, the 100 metre tall tower of Golm transmitter near Potsdam, Germany was built from recycled timber. It stood for 31 years. It was common to reuse wood of dismantled radio towers in the 1930s in Germany, e.g. the former tower of Koblenz radio transmitter was built of wood from a dismantled tower, which carried a T-antenna at Transmitter Muehlacker. The upper parts of the 157-metre-tall wood tower of Ismaning radio transmitter, which stood for 49 years, were built of wood from a smaller radio tower dismantled in 1934.

Reclaimed Lumber

A lounge chair using reclaimed wood.

Reclaimed lumber is processed wood retrieved from its original application for purposes of subsequent use. Most reclaimed lumber comes from timbers and decking rescued from old barns, factories and warehouses, although some companies use wood from less traditional structures such as boxcars, coal mines and wine barrels. Reclaimed or antique lumber is used primarily for decoration and home building, for example for siding, architectural details, cabinetry, furniture and flooring.

This dining hall uses wood recycled from barns for flooring, walls, and furniture.

In the United States of America, wood once functioned as the primary building material because it was strong, relatively inexpensive and abundant. Today, many of the woods that were once plentiful are only available in large quantities through reclamation. One common reclaimed wood, longleaf pine, was used to build factories and warehouses during the Industrial Revolution. The trees were slow-growing (taking 200 to 400 years to mature), tall, straight, and had a natural ability to resist mold and insects. They were also abundant. Longleaf pine grew in thick forests that spanned over 140,000 square miles (360,000 km²) of North America. Reclaimed longleaf pine is often sold as Heart Pine, where the word "heart" refers to the heartwood of the tree.

Previously common woods for building barns and other structures were redwood (Sequoia sempervirens) on the U.S. west coast and American Chestnut on the U.S. east coast. Beginning in 1904, a chestnut blight spread across the US, killing billions of American Chestnuts, so when these structures were later dismantled, they were a welcome source of this desirable but later rare wood for subsequent reuse. American Chestnut wood can be identified as pre- or post-blight by analysis of worm tracks in sawn timber. The presence of worm tracks suggests the trees were felled as dead standing timber, and may be post-blight lumber.

Barns are one of the most common sources for reclaimed wood in the United States. Those constructed through the early 19th century were typically built using whatever trees were growing on or near the builder's property. They often contain a mix of oak, chestnut, poplar, hickory and pine timber. Beam sizes were limited to what could be moved by man and horse. The wood was often hand-hewn with an axe and/or adze. Early settlers likely recognized American oak from their experience with its European species. Red, white, black, scarlet, willow, post, and pin oak varieties have all been used in North American barns.

Mill buildings throughout the Northeast also provide an abundant source of reclaimed wood. Wood that is reclaimed from these buildings includes structural timbers - such as beams, posts, and joists - along with decking, flooring, and sheathing. These buildings often have no economic or reuse possibility, can be a fire hazard, and may require varying degrees of environmental cleanup. Reclaiming lumber and brick from these retired mills is considered a better use of materials than landfill-based disposal.

Another source of reclaimed wood is old snowfence. At the end of their tenure on the mountains and plains of the Rocky Mountain region, snowfence boards are a valued source of consistent, structurally sound and reliable reclaimed wood. Other woods recycled and reprocessed into new wood products include coast redwood, hard maple, Douglas Fir, walnuts, hickories, red and White Oak, and Eastern white pine.

Properties

Reclaimed lumber is popular for many reasons: the wood's unique appearance, its contribution to green building, the history of the wood's origins, and the wood's physical characteristics such as strength, stability and durability. The increased strength of reclaimed wood is often attributed to the wood often having been harvested from virgin growth timber, which generally grew more slowly, producing a denser grain.

Reclaimed beams can often be sawn into wider planks than newly harvested lumber, and many companies claim their products are more stable than newly-cut wood because reclaimed wood has been exposed to changes in humidity for far longer.

Reclaimed Lumber Industry

The reclaimed lumber industry gained momentum in the early 1980s on the West Coast when large-scale reuse of softwoods began. The industry grew due to a growing concern for environmental impact as well as declining quality in new lumber. On the East Coast, industry pioneers began selling reclaimed wood in the early 1970s but the industry stayed mostly small until the 1990s as waste disposal increased and deconstruction became a more economically alternative to demolition. A trade association, the Reclaimed Wood Council, was formed in May 2003 but dissolved in January 2008 due to a lack of participation among the larger reclaimed wood distributors.

Reclaimed lumber is sold under a number of names, such as antique lumber, distressed lumber, recovered lumber, upcycled lumber, and others. It is often confused with salvage logging.

LEED

The Leadership in Energy and Environmental Design (LEED) Green Building Rating System is the US Green Building Council's (USGBC) benchmark for designing, building and operating green buildings. To be certified, projects must first meet the prerequisites designated by the USGBC and then earn a certain number of credits within six categories: sustainable sites, water efficiency, energy and atmosphere, materials and resources, indoor environmental quality, innovation and design process.

Using reclaimed wood can earn credits towards achieving LEED project certification. Because reclaimed wood is considered recycled content, it meets the 'materials and resources' criteria for LEED certification, and because some reclaimed lumber products are Forest Stewardship Council (FSC) certified, they can qualify for LEED credits under the 'certified wood' category.

Drawbacks

With reclaimed material being so popular, it is becoming more difficult to source. With such a high demand, some sellers try to pass newer wood off as antique. It is also common (although not necessarily done intentionally) for species to be misidentified because it is difficult to tell the difference in older material unless it is cut open and examined, leaving the material less desirable. Professionals in the field and with established reclaimed wood enterprises do not have difficulty identifying the species.

Reclaimed lumber sometimes has pieces of metal embedded in it, such as broken off nails, so milling the material can often ruin planer knives, saw blades, and moulder knives. Nail-compatible saw blades are advisable for the same reason, as well as for safety. The alternative is to remove all metal from the reclaimed lumber, which is a costly and tedious process commonly achieved by scanning each piece of wood with a metal detector and then manually pulling out all nails, bolts, bullets, screws, buckshot, and other miscellaneous metal hardware. This process can make the cost of reclaimed lumber higher than new lumber.

Many sources of reclaimed wood cannot verify what the wood might have been treated with over its lifetime. This uncertainty leads to fears of harmful offgassing of volatile organic compounds associated with lead paint or various stains and treatments that may have been used on the wood. These fears are particularly pressing when the wood is for an interior application.

PAINT RECYCLING

Paint is a recyclable item. Latex paint is collected at collection facilities in many countries and shipped to paint-recycling facilities.

How Paint is Recycled?

There are many ways that paint can be recycled. Most often, the highest quality of latex paint is sorted out and turned back into recycled paint that can be used. Recycled paint is environmentally preferable to new paint, while still maintaining comparable quality. In many cases, reusable paints of the same color are pumped into a tank where the material is mixed and tested. The paint is adjusted with additives and colorants as necessary. Finally, the paint is fine filtered and packaged for sale.

Paint that cannot be reused has other environmentally friendly uses. Non-reusable paint can be made into a product used in cement manufacturing, thereby recycling virtually 100% of the original paint.

Recycling one gallon of paint could save 13 gallons of water, 1 quart of oil, and 250,000 gallons of water pollution, 13.74 pounds of CO_2, save enough energy to power the average home for 3 hours, or cook 6 meals in a microwave oven, or blow dry someone's hair 27 times.

Paint Recycling by Country

In Ontario, Stewardship Ontario oversees the collection of waste paint from consumers and diversion from landfill to meet targets approved by the Ministry of the Environment through a program called the Orange Drop Program. The Orange Drop program is an extensive and growing network of collection sites—drop-off locations for paint leftovers and other special materials that can't go in the Blue Box or the garbage.

As an Orange Drop-approved transporter and processor, Loop Recycled Products Inc. takes leftover paint, collected through Stewardship Ontario, and turns it into 12 shades of premium, affordable and environmentally friendly recycled paint. Reusing top-quality residual paint (on average, the original retail value of a gallon of incoming paint is approximately $30) enables Loop to create premium products without the raw material costs and energy consumption needed to make paint from scratch.

Since 2012, Loop Recycled Products Inc. has diverted over 6 million litres of paint from disposal in Ontario's landfills, incineration and waterways and is committed to innovation and solving Canada's waste paint problem. In February 2015 Waste Diversion Ontario approved Product Care as the new Ontario waste paint stewardship operator effectively replacing Stewardship Ontario.

In March 2017, Colortech ECO Paints introduced its line of recycled wall and floor paints to specific retail markets consisting of a large network of liquidation and discount stores across Canada and the United States., as well as exporting large quantities to West Africa and South America.

Alberta's paint recycling program started accepting leftover, unwanted paint on April 1, 2008. It is estimated that about 30 million liters of paint is sold in Alberta each year. On average, 5 to 10 percent of this ends up as waste, which can pose environmental and health risks if disposed of improperly. Paint contains many components that have great potential for reuse, recycling and recovery. The Paint Recycling Alberta program enables these products to be handled and recycled in an environmentally safe manner, reducing their impact on the environment. The program is funded through environmental fees charged on the sale of new paint in Alberta. The fees are put into a dedicated fund that can only be used to manage the paint recycling program.

The paint is sorted into different streams and sent to registered processors to be recycled into new paint, used in other products or in energy recovery, or sent for proper disposal if necessary. Any processor that receives paint must be registered with the Paint Recycling Program and meet all applicable environmental, transportation, health & safety, and local requirements.

Calibre Environmental LTD. (CEL) located in Calgary, Alberta, became a key part in 2008 of the new Alberta Paint Stewardship program which significantly increased the recycling of unused latex paint from across the province of Alberta. Calibre Environmental Ltd. currently processes about 1.6 million kilograms of latex paint annually, which equates to the successful recycling of one million litres of quality latex paint per year.

In the UK reusable leftover paint can be donated to Community RePaint, a national network of paint reuse schemes. The network comprises local schemes run by not-for-profit organisations, local authorities or waste management companies, in the Community RePaint network. The schemes collect surplus paint from trade sources i.e. painters, decorators, retailers, manufacturers, and/or leftover paint donated by householders at council household waste and recycling centres (also known as tips). The paint is then sorted by staff and volunteers before being redistributed to local charities, community groups, families and individuals in need. The Community RePaint network, is sponsored by Dulux (part of AkzoNobel), managed by an environmental consultancy, Resource Futures and has been cited as an example of best practice for the management of surplus paint in a report by the European Commission and by DEFRA in Guidance on Applying the Waste Hierarchy.

There are also a handful of companies recycling paint in the UK, but only two reprocessing waste paint, back into the quality of virgin paint; Newlife Paints. Newlife Paints was formed in 2008 after Keith Harrison, an industrial chemist, developed a process that converted waste emulsion paint back into full quality, commercial grade paint. Castle RePaint, part of the social enterprise company Castle Furniture also consolidates unwanted emulsion paint into brand new 'RePaint' in a range of colours.

Concerns about the life cycle of paint have led to the creation of PaintCare, a non-profit 501(c)(3) organization established to represent paint manufacturers (paint producers) to plan and operate paint stewardship programs in the United States in those states that pass paint stewardship laws.

Paint stewardship law aims to enable the paint industry to implement a collection program that allows consumers to take their leftover, unwanted paint to a collection site to be collected and recycled. Legislation mandating the creation of the PaintCare program has been enacted in eight states since 2009: Oregon, California, Connecticut, Rhode Island, Vermont, Minnesota, Maine, and Colorado. Legislation has also been passed for the District of Columbia; PaintCare anticipates beginning the District's paint stewardship program in September 2016. PaintCare is responsible for promoting the reuse of post-consumer architectural paint (leftover paint) and providing for the collection, transport, and processing of this paint using the hierarchy of "reduce, reuse, recycle," and proper disposal. Most PaintCare locations are at paint retailers who volunteer to take back paint. These retailers take back paint during regular business hours, making paint recycling and disposal much more convenient for the public.

Paint is shipped to companies such as GDB International, Amazon Paint, American Paint Recyclers (Ohio), Metro Paint (Oregon), UCI Environmental (Nevada) and Kelly Moore, Visions Paint Recycling, Inc (California)& Williams Paint Recycling Company. In the Southern California area, Acrylatex Coatings & Recycling, Inc. accepts unused/unwanted latex paints for reprocessing into a viable resource of recycled paints in 20-standard colors. In the southeastern United States Atlanta Paint Disposal has a paint recycling program with drop off locations in Atlanta, Georgia. In the northeast The Paint Exchange, LLC recycles latex paint under the brand recolor. In the Mid-Atlantic RepaintUSA also recycles paint with a focus on businesses, public and private sources.

A new charitable organization known as The Global Paint for Charity incorporated in Georgia, US, has as its mission to collect leftover paint from residents and businesses nationwide and use it for global housing rehabilitation projects, including homes, schools, hospitals, jails and churches for vulnerable families in developing countries.

They partner with non-profit organizations with existing operations in these continents for paint distribution. Through the support of their donors and partners they are able to improve communities, increase access to quality paints and protect the environment.

The Environmental Protection Agency (EPA) estimates that every homeowner in the US has 3 to 4 gallons of leftover paints in their basement, and 10 percent of those paints ends up in landfills. One gallon of improperly disposed paint has the ability to pollute up to 250,000 gallons of water.

By participating in the program, individuals and businesses will take greater steps to protect the environment, and improve living conditions for vulnerable populations throughout the world. If you would like to support the Global Paint for Charity, they encourage you to take action today.

Improve access of high-quality paints to vulnerable populations around the world. Nearly 2.5 billion people in developing countries live on less than $2 a day. For them, paint is very expensive. In these settings, it is very difficult for families to secure sufficient income for their basic needs (including but not limited to: Food, medicine, water, clothes, school supplies, and shelter). When making consumption choices that involve spending on their basic needs there is nothing left to spend on paint. The paint shortage affects many other areas of the world, where communities lack even the most basic need and materials to uplift their people. For the world's poorest communities, home isn't just where the heart is. Dirt walls and neglected communities are not attractive to tourists, putting those who cannot afford paint, not only at the greatest risk of life-threatening of bad germs but also lack of economic opportunities.

Since it started years ago, as many as 500–6000 gallons of paint have been shipped at a time to developing countries, including Kenya, Uganda, Liberia, Benin, Cameroon, Ghana, Guinea, Bahamas, Panama, Haiti, Dominican Republic, Honduras, El Salvador, Guyana, Jamaica and Mexico, 890 volunteers have painted 10,000 family homes and 320 schools and orphanages with over 280,000 tons of donated paint from businesses and residences.

Global Paint for Charity continues to expand and impact our communities around the world. They do this through working directly with their volunteers, donors, and partners. From developing paint projects that engage their employees in the beautification of the community; to run frequent local paint drives to support their program.

Global Paint for Charity has recently won the National Energy Globe Award of United States 2017. With more than 178 participating countries and over 2,000 project submissions annually the Energy Globe Award is today's most prestigious environmental prize worldwide. The Austrian Honorary Consul General, Mr. Ferdinand C. Seefried hosted an exclusive ceremony to present the National Energy Globe Award 2017 for the United States to Mr. Rony Delgarde, Global Paint for Charity.

VEGETABLE OIL RECYCLING

A collection drum for used cooking oil, as an input to creating a form of biodiesel that modified diesel engines can run on.

Vegetable oil recycling is increasingly being carried out to produce a vegetable oil fuel. In the UK, waste cooking oil collection is governed by the Environment Agency. All waste cooking oil collections need to be carried out by a company registered as a waste carrier by the Environment Agency. On each collection a waste transfer note needs to the filled out and copies held by both parties for a minimum of 3 years. Waste transfer notes need to contain:

- Full company details of who the waste is being transferred to,
- Their waste registration details,
- Full details of who the waste is being transferred from,
- Date,
- Signatures from both parties.

Waste transfer notes can be hard paper copies or electronic versions. Here is an example of a waste transfer note currently is use by a UK waste cooking oil company. Opportunities for businesses and consumers to recycle used cooking oil ("yellow grease") has increased. Used cooking oil can be refined into different types of biofuels used for power generation and heating. A significant benefit is that biofuels derived from recycled cooking oil typically burn clean, have a low carbon content and do not produce carbon monoxide. This helps communities to reduce their carbon footprints. The recycling of cooking oil also provides a form of revenue for restaurants, which are sometimes compensated by cooking oil recyclers for their used deep fryer oil. Cooking oil recycling also results in less used oil being disposed of in drains, which can clog sewage lines due to the build-up of fats and has to be collected there as "brown grease" by grease traps.

Vegetable oil refining is a process to transform vegetable oil into fuel by hydrocracking or hydrogenation. Hydrocracking breaks larger molecules into smaller ones using hydrogen while hydrogenation adds hydrogen to molecules. These methods can be used for production of gasoline, diesel, and propane. The diesel fuel that is produced has various names including green diesel or renewable diesel.

In the past waste oils were collected by pig farmers as part of food waste from pig swill bins. The grease was skimmed off the swill tanks and sold for further processing, while the remaining swill was processed into pig food.

WATER RECYCLING

Water recycling is the process of treating waste water and reusing it. Recycled water can be reused for the same process, for irrigation or as an alternative to mains water in wash-down applications. Water recycling systems will vary according to the quality of waste water to be treated and the intended application for the water. The process may involve the use of an oil and water separator, a filtration system, a detergent removal unit and a sanitation unit.

Key Features of Cleanawater Water Recycling Systems

- AQIS compliant technology.

- Retrofit wash bays for AQIS compliant upgrades.

- Local Water Authority compliant systems.

- Track flow rates, pH levels, usage statistics and storage levels.

- Equipped with controllers, alarms, sensors and switches to alert you immediately of faults.

Industry Examples

Example: Car Wash.

A self-service car wash is looking to implement a water recycling system. The business primarily wants to reduce mains water expenses and waste water disposal fees. However, they also recognise the potential marketing advantage of becoming a green and environmentally friendly business.

The car wash operates 24/7 all year around without supervision from operators. It also uses a range of fast breaking detergents as part of the washing process. A solution will need to have a remote alert system, as well as the ability to treat the various detergents and chemicals in the waste water.

The separation of the fast breaking detergents occurs in the triple interceptor. Here the detergents are given time to break, separate, and sink to the bottom. Once detergent has been removed, water is pumped through an oil and water separator where oil and large particles are removed from the water into a waste tank.

The waste water then travels to a filtration unit that filters particles down to 1 micron in size. Upon completing this process, water enters the detergent removal unit where detergents, shampoo, waxes and other cleaning agents are removed. Finally, water is sterilised in order to make it safe to be used around humans.

Example: Importing goods.

A small international trade business is planning to import goods into Australia from a number of other countries. The goods will be imported in containers and wooden pallets that will need to be washed down according to the Australian Quarantine and Inspection Service (AQIS) and the Department of Agriculture, Fisheries and Forestry (DAFF).

The business will need an AQIS wash down bay so that the goods can be washed down to remove any soil, insects, seeds and other particles that pose as a threat. An AQIS water recycling system will also need to be installed so that water is treated to both Australian Quarantine and local trade waste standards.

AQIS has strict policies for wash down bay requirements that include splash walls' height, and cleaning procedures for wash pads and protective clothing.

Some of the DAFF standards for water recycling systems include:

- Filtering organic contaminants to a minimum of 100 microns.

- A post-filtration pH between 5.0 and 7.0.

- The mechanical agitation of water post-chlorination.

- Chlorine levels of at least 200 parts per million.

How a Water Recycling System Works?

The Need for a Customised Solution

Water recycling systems work by taking waste water and treating it until it is suitable for reuse in the intended application. Water recycling systems vary depending on the type of water to be recycled and the requirements of intended application.

Industries that use vast amounts of water will benefit from water recycling systems used in their day to day operation. Depending on the quality of waste water, the number of steps in water recycling system will vary.

In the car wash industry for example, waste water will contain cleaning agents, oil, food scraps and other debris which wash into the pit that collects waste water. The availability of space to install the water recycling system, and the required flow rate are other contributing factors when designing the best water recycling system for your application.

Water Recycling Stages

Removing Oil and Large Particles

The first step of water recycling is to remove oil and large particles from the water. This first step is possible thanks to the combined efforts of the triple interceptor and oil separator.

A triple interceptor is a three-compartment tank where water overflows from the first compartment to the second, and then from the second to the third. The main purpose of a triple interceptor is to allow time for any sludge to sink to bottom of the first compartment.

Once waste water has gone through a triple interceptor, it enters the oil and water separator. The fundamental principle behind a water and oil separator is simple physics: since oil is less dense than water it will tend to float to the top of a tank, where it can be collected and removed.

Basic oil and water separators can be broadly divided into three groups:

- Separators that primarily rely on gravity for the separation of oil to occur.

- Separators that provide a medium on which oil particles can coalesce.

- Separators which apply high centrifugal forces to waste water to separate oil particles.

Cleanawater's expert team will help you determine which oil and water separator is the best fit to your business and waste water requirements.

Filtration

After oil and large solids have been removed, the waste water is deposited into a process tank where it is stored before being pumped into a filtration system tank. The first stage in the filtration system is a deep bed media filter. This is where water is pushed at high pressure through fine sand and other granular particles in order to remove any large particles still present in the water. To ensure that the required maximum particle size is achieved, water is then put through 3 stage cartridge filtration capable of reducing sediment size to 1 micron.

Removing Chemicals

The next step in the water recycling system depends on the presence, type and amount of surfactant in the water. Surfactants are commonly found in detergents. They help wet surfaces more thoroughly by reducing water surface tension.

Detergents work by allowing soil and oily residue to mix with water and be rinsed off easily. The effectiveness of detergents and soaps depends on a number of factors, including water hardness. Water hardness is the measure of calcium, magnesium, iron and manganese ions in water. Calcium and magnesium ions bind to the surfactant and leave an unpleasant residue on water and on surfaces.

To address this residue, substances called builders are added to the water to remove these problematic ions, thus making water soft. The most common builder added to surfactants is phosphate. While phosphate solves the water hardness issue, it creates another problem -- phosphate and surfactants don't degrade in the environment, and are highly toxic for animals and vegetation.

To respond to the need for environmentally friendly detergents, manufacturers have created fast breaking detergents. These detergents lower the water surface tension for a period of time, which allows water and soil to emulsify. After some time has passed, the surface tension returns to normal allowing the separation of water and soil. Biodegradable detergents, which can be broken down by bacteria, have also been introduced.

There are two main ways to remove detergent from water:

- The use of specific chemicals which allow detergent particles join and coagulate.

- Treating water in a tank using bacteria to decompose detergent.

Sterilisation

Once water is free of detergents, a sterilisation phase can commence. Waste water is full of pathogens which are disease-causing organisms, viruses, parasites and bacteria that are found in poor quality water.

Although this stage is only required for when recycled water is to be used around humans, it is highly recommended as a standard. Pathogens can cause severe illness if infected water comes in contact with humans.

In a car wash setting for instance, water is sprayed through high pressure hoses to maximise water saving, but this means that mist can travel long distances and the risk of harmful water coming in contact with a person increases.

There are two ways of sterilising water:

- Chlorine sterilisation is achieved by adding and maintaining a concentration of Chlorine to the water. It is ideal for when water is to be stored and reused in the future.

- UV steralisation is achieved by running water under UV light rays. This disrupts pathogens' cellular functioning and is ideal for when water is to be used immediately.

References

- Recycling-thermoset-composites, international-composites-news: jeccomposites.com, Retrieved 23 May, 2020

- Müller, Lothar (2014). White Magic: The Age of Paper. Polity Press. ISBN 978-0-7456-7253-3. Retrieved 28 December 2019

- How-water-recycling-systems-work, information-centre: cleanawater.com.au, Retrieved 2 May, 2020

- Pivnenko, K.; Pedersen, G. A.; Eriksson, E.; Astrup, T. F. (1 October 2015). "Bisphenol A and its structural analogues in household waste paper". Waste Management. 44: 39–47. doi:10.1016/j.wasman.2015.07.017. PMID 26194879

- The-importance-of-copper-recycling-2877931: thebalancesmb.com, Retrieved 21 February, 2020

3

Product Recycling

Products made from a variety of materials can be recycled using a number of processes. It includes products and items like cardboard, tire, bottle, lamps, textiles, batteries, oil filters, etc. This chapter closely examines product recycling to provide an extensive understanding of the subject.

CARDBOARD RECYCLING

Cardboard, also referred to as corrugated cardboard, is a recyclable material that is recycled by small and large scale businesses to save money on waste disposal costs. Cardboard recycling is the reprocessing and reuse of thick sheets or stiff multilayered papers that have been used, discarded or regarded as waste. Cardboard boxes are usually heavy-duty or thick-sheets of paper known for their durability and hardness. Examples of cardboard include packaging boxes, egg cartons, shoe boxes, and cereal boxes.

Recycling is good for us as it not only saves our environment from deterioration by reducing pollution but also conserves valuable resources and creates jobs. Cardboard recycling is done as a way of keeping the environment clean and green. The steps below provide an explanation of the cardboard recycling system.

Step-by-Step Process of Cardboard Recycling

Collection

"Collection" is the first step in recycling cardboard. Recyclers and businesses collect the waste cardboard at designated cardboard collection points. Majority of the collection points include trash bins, stores, scrap yards, and commercial outlets that generate cardboard waste. After collection, they are then measured and hauled to recycling facilities, mostly paper mills.

At this point, there are certain types of cardboard that are accepted while some are not depending on how they were used or manufactured. For instance, cardboard that is waxed and coated or used for food packaging are not accepted in most cases as they undergo different specialized recycling process.

Sorting

Once the corrugated boxes arrive at the recycling facility, they are sorted according to the materials they are made of. In most cases, they are classified into corrugated cardboard and boxboard. Boxboards are the ones that are thin such as those used for cardboard drink containers or cereals boxes while corrugated boxes are bigger and stiffer commonly used for packaging transport goods. Sorting is important since paper mills manufacture different grades of materials based on the materials being recovered.

Shredding and Pulping

After sorting is done, the next step is shredding then pulping follows. Shredding is done to break down the cardboard paper fibers into minute pieces. Once the material is finely shredded into pieces, it is mixed with water and chemicals to breakdown the paper fibers that turn it into a slurry substance.

This process is what is termed as pulping. The pulped material is then blended with new pulp, generally from wood chips that ultimately help the resulting substance to solidify and become firmer.

Filtering, Conterminal Removal and De-inking

The pulp material is then taken through a comprehensive filtering process to get rid of all the foreign materials present as well as impurities such as strings, tape or glue. The pulp further goes into a chamber where contaminants like plastics and metals staples are removed through a centrifuge-like process. Plastics float on top while the heavy metal staples fall to the bottom after which they are eliminated.

The next process, de-inking, involves putting the pulp in a floatation device made up of chemicals that take away any form of dyes or ink via a series of filtering and screening.

This step is also called the cleaning process as it cleans the pulp thoroughly to ensure it is ready for the final processing stage.

Finishing for Reuse

At this stage, the cleaned pulp is blended with new production materials after which, it is put to dry on a flat conveyor belt and heated cylindrical surfaces. As the pulp dries, it is passed through an automated machine that press out excess water and facilitates the formation of long rolls of a solid sheet from the fibers called linerboards and mediums. The linerboards are glued together, layer by layer to make a new piece of cardboard.

In other cases, the medium is used as the corrugated sheet which is taken through two huge metal rolls with teeth to give it the ridges. Linerboards are then glued to the medium as the thin outer covering. Alternatively, the linerboards and mediums are ferried to boxboard manufacturers where the manufacturing process is completed by use of machines that shape and create a crease along pattern folds to make the boxes used for packaging or transporting products.

Advantages of Cardboard Recycling

Reuse and Environmental Conservation

Cardboard recycling helps to reduce the dumping of cardboard waste in landfills. On this basis, cardboard recycling helps to improve the cleanliness of the environment and promotes healthy human surrounding. Some cardboard is made from almost 100% recycled materials, while the majority are averaged at 70% to 90%.

The cardboard materials are also biodegradable thus, it reduces environmental footprint and degradation by offering what is termed as "green" packaging solutions. Recycling cardboard also offers a fundamental solution to environmental conservation by preserving natural resources due to its highest reuse percentage in producing new cardboard products.

Promotes the use of Renewable Materials

The wood chips materials added during pulping are made from birch or pine tree pulp that have a high percentage of recyclable content. Furthermore, these trees are easy to grow in various environmental conditions. They are also first growing compared to hardwood trees. Due to the fact that they are able to grow fast in various conditions and their recyclable quality, it means they can be managed and harvested sustainably thereby promoting the use of renewable materials.

Saves Energy

Owing to the highest percentage of cardboard recyclability, the amount of energy required for producing corrugated packing products is tremendously reduced. Further, they are made from locally available materials that can be harvested in an environmentally friendly way.

On this basis, the transportation and production costs of cardboard recycling are reduced while at the same time excelling at delivering materials of greater structural strength for packaging or protecting goods in transit. This means, no extraneous materials and energy are needed to manufacture new cardboard boxes. As a result, recycling cardboard saves energy and materials needed to make new cardboard and also reduces environmental pollution.

Why Recycle Cardboard?

Recycling is good for our planet as it helps to conserve resources, create jobs and reduces pollution from the production of new materials. Big businesses also recycle items as part of their corporate social responsibility programs and it also helps them to save money on waste disposal costs.

Many of us buy so many items online, we have lots of cardboard boxes to dispose of. There are several ways to dispose of these items like:

- Re-use: When you have plenty of cardboard boxes at your disposal, just use them for storage purposes. You can even make toys for your kids from large cardboard size boxes. You can also use them as containers to separate different items for recycling purposes.

- Pass them: You can pass on big cardboard size boxes to your friends or relatives in case they are planning to shift home. Small cardboard boxes can be used to store items that are not used too frequently. School children can use these cardboard boxes for their school projects.

- Recycle: All those pieces of cardboard boxes that you are not able to re-use for any reason, you may send them to the local recycling center. Cardboard is accepted at most of the waste recycling centers.

AUTOMOTIVE OIL RECYCLING

Automotive oil recycling involves the recycling of used oils and the creation of new products from the recycled oils, and includes the recycling of motor oil and hydraulic oil. Oil recycling also benefits the environment: increased opportunities for consumers to recycle oil lessens the likelihood of used oil being dumped on lands and in waterways. For example, one gallon of motor oil dumped into waterways has the potential to pollute one million gallons of water.

Waste oil collection for recycling at the Fairgreen Amenity Site, Portadown.

Motor Oil

Oil being drained from an automobile.

Recycled motor oil can be combusted as fuel, usually in plant boilers, space heaters, or industrial heating applications such as blast furnaces and cement kilns. When used motor oil is burned as fuel it must be burned at high temperatures to avoid gaseous pollution. Alternatively, waste motor oil can be distilled into diesel fuel or marine fuel in a process similar to oil re-refining, but without the final hydrotreating process. The lubrication properties of motor oil persist, even in used oil, and it can be recycled indefinitely.

Used Motor Oil Re-refining

Used oil re-refining is the process of restoring used oil to new oil by removing chemical impurities, heavy metals and dirt. Used industrial and automotive oil is recycled at re-refineries. The used oil is first tested to determine suitability for re-refining, after which it is dehydrated and the water distillate is treated before being released into the environment. Dehydrating also removes the residual light fuel that can be used to power the refinery, and additionally captures ethylene glycol for re-use in recycled antifreeze.

Next, industrial fuel is separated out of the used oil then vacuum distillation removes the lube cut (that is, the fraction suitable for reuse as lubricating oil) leaving a heavy oil that contains the used oil's additives and other by-products such as asphalt extender. The lube cut next undergoes hydro treating, or catalytic hydrogenation to remove residual polymers and other chemical compounds, and saturate carbon chains with hydrogen for greater stability.

Final oil separation, or fractionating, separates the oil into three different oil grades: Light viscosity lubricants suitable for general lubricant applications, low viscosity lubricants for automotive and industrial applications, and high viscosity lubricants for heavy-duty applications. The oil that is produced in this step is referred to as re-refined base oil (RRBL).

The final step is blending additives into these three grades of oil products to produce final products with the right detergent and anti-friction qualities. Then each product is tested again for quality and purity before being released for sale to the public.

But you can not simply compare those ratios and conclude that refining from crude is immensely inefficient. Crude oil refining yields large amounts of fuels. Below is a comparison of refining from used motor oil and refining from crude.

Re-refining one unit of used motor oil will yield:

- 71% lube oil.

- 5% fuels.

- 14% asphalt.

- 10% water.

Refining one unit of crude oil will yield:

- 84% fuels (46% gasoline, 38% other fuels).

- 9% gases.

- 4% coke.

- 3% asphalt and road oil.

- 3% petrochemical feed-stocks.

- 1% lubricating oil.

REOB

The sludge ("residue") associated with engine oil recycling, which collects at the bottom of re-refining vacuum distillation towers, is known by various names, including "re-refined engine oil bottoms" (abbreviated "REOB" or "REOBs").

A report from the U.S. Federal Highway Administration (FHWA) states that: The oil in a car engine contains a variety of additives to enhance the vehicle's performance. These include polymers, viscosity modifiers, heat stabilizers, additional lubricants, and wear additives. The REOB contains all the additives that were in the waste oil as well as the wear metals from the engine (mainly iron and copper). These additives include zinc dialkyldithiophosphate, which contains zinc, sulfur, and phosphorus; calcium phenate, which contains calcium; and molybdenum disulfide, which contains molybdenum and sulfur.

Some producers of asphalt for paving have—openly or secretly—incorporated REOBs into their asphalt, creating some controversy and concern in the traffic engineering community, with some experts suggesting it reduces the durability of the resulting pavement.

Used oil, or 'sump oil' as it is sometimes called, should not be thrown away. Although it gets dirty, used oil can be cleaned of contaminants so it can be recycled again and again. There are many uses for recycled used oil.

These include:

- Industrial burner oil, where the used oil is dewatered, filtered and demineralised for use in industrial burners.

- Mould oil to help release products from their moulds (e.g. pressed metal products, concrete).

- Hydraulic oil.

- Bitumen based products.

- An additive in manufactured products.

- Re-refined base oil for use as a lubricant, hydraulic or transformer oil.

Once you have taken your used oil to your local collection facility, used oil collectors take the used oil and undertake some pre-treatment and recycling of the used oil or sell it to a specialised used oil recycler.

Pre-treatment or Dewatering

Pre-treatment of used oil involves removing any water within the oil, known as dewatering. One way of doing this is by placing it in large settling tanks, which separates the oil and water.

Further recycling steps include:

- Filtering & demineralisation of the oil, to remove any solids, inorganic material and certain additives present in the oil, producing a cleaner burner fuel or feed oil for further refining.

- Propane de-asphalting to remove the heavier bituminous fractions, producing re-refined base oil.

- Distillation to produce re-refined base oil suitable for use as a lubricant, hydraulic or transformer oil. This process is very similar to the process undergone by virgin oil.

Water is found in used oil as free water or bound water, for example in emulsions. The term dewatering is usually taken to mean the removal of free water. Where water has been emulsified with oil, the emulsion has to be "broken" with a demulsifier before the water can be separated from the oil.

Dewatering is a simple process relying on the separation of aqueous and oil phases over time under the influence of gravity. The used oil is allowed to stand in a tank (raw waste oil) and free water drops to the bottom where it can be drained, treated (waste water treatment) and discharged appropriately to sewer or stormwater depending on quality and local regulations.

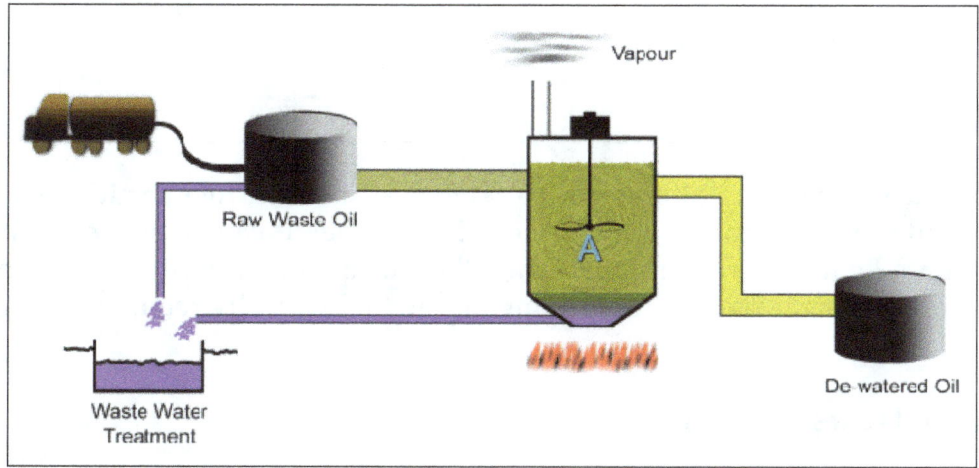

Heating and stirring the used oil in a tank (A) and driving off the water through evaporation can speed up the dewatering process.

The "dried" or dehydrated oil is then suitable for further processing or for use as a burner fuel.

Filtering and Demineralisation

The purpose of filtering and demineralisation is to remove inorganic materials and certain additives from used oil to produce a cleaner burner fuel or feed for re-refining. Used oil feedstock is transferred to a reaction tank (A) and mixed with a small quantity of sulphuric acid and heated to about 60oC. A chemical surface-active reagent, called a surfactant, is added to the reactor (A) and after stirring the mixture is allowed to stand. This allows the mixture to separate into two "phases" - i.e. oil and water-based or aqueous. The reagent causes the contaminants to accumulate in the aqueous phase, which settles to the bottom of the tank (A) and is drained off as slurry. This phase contains acid, used oil contaminants, including metals and some of the oil additives. The water is dried off, leaving a solid waste that must be disposed of.

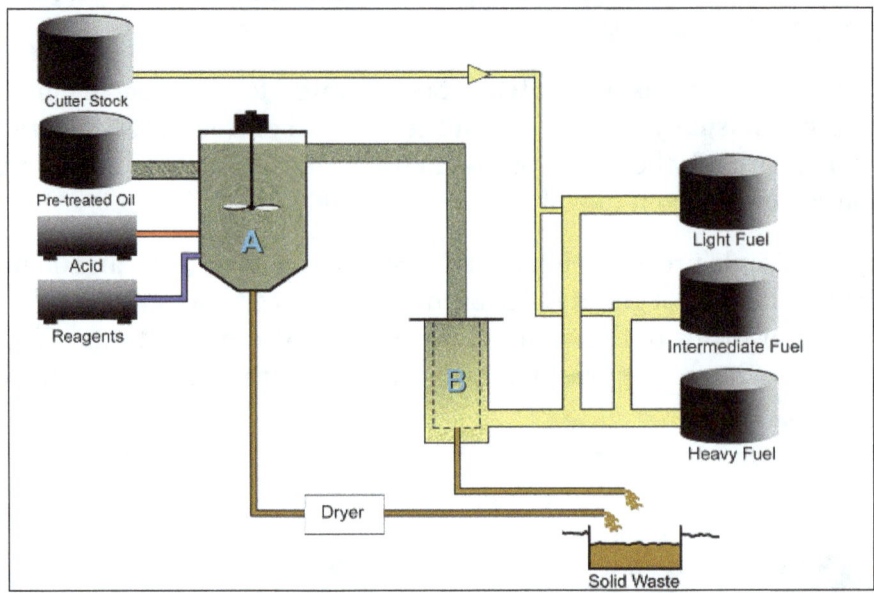

The demineralised oil is filtered (B) to remove suspended fine particles (to solid waste) and run off to storage (C) as a clean burner fuel. It can be further diluted or "cut" with a lighter petroleum product (called cutter stock) to produce a range of intermediate to light fuel oils depending on the fuel viscosity requirements of the burner.

Propane De-asphalting

The Propane De-asphalting (PDA) process is an important pre-treatment step in the re-refining process producing de-asphalted lube-oil, which becomes a feedstock for the next step in a re-refining facility. The other output (which is also an input) is propane, which is recovered from both streams and re-used within the process.

The PDA process relies on the greater solubility of the paraffinic and naphthenic (ie essentially the base oil) components versus the contaminated waste material in a stream of propane.

The separation of the lubricating oil fraction from used oil is a continuous process and is conducted at ambient temperature when processing used oil.

The used oil is pumped into the middle of the extraction column (A). Liquid propane is charged to the bottom of the column (A). The oil being heavier than propane, flows down the column (A); the propane rises in a counter-flow thus mixing the input streams within the column (A). The rising propane dissolves the more soluble lube oil components, which are carried out the top of the column (A) with the propane, and the propane insoluble material is removed from the bottom of the column (A).

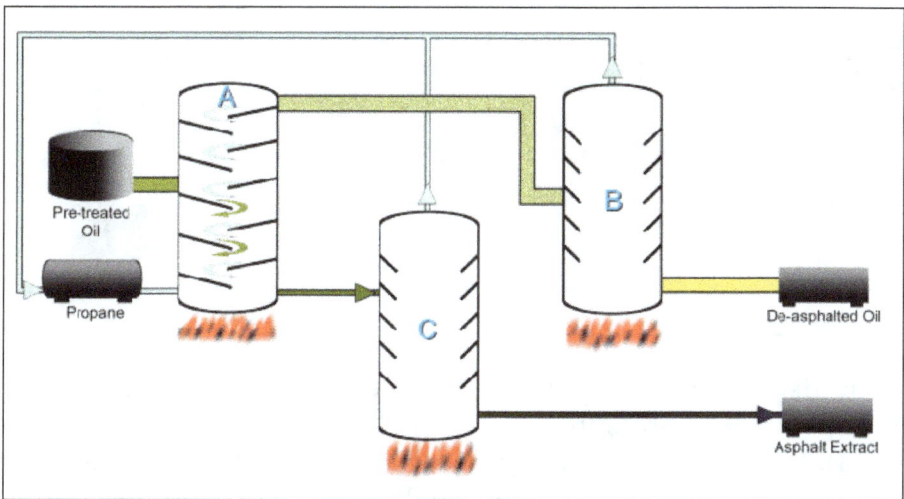

Propane is vaporised from both streams [ie., the de-asphalted lube-oil stream (B) and the waste stream (C)] in "stripper" units (B) and (C), then condensed and returned to the propane storage tank.

The de-asphalted lube-oil component is feed for the next processing stage. The residuum (waste) component is mixed with bottoms from the vacuum distillation tower to produce an asphaltic material.

Distillation

Distillation (or Fractionation) is the physical separation of components of lubricating oil by boiling range. Depending on the type of distillation, the boiling ranges can produce gases and gasolines at the lower boiling points with heavy lubricating oils being distilled at higher boiling points. Distillation is the core process for a facility capable of producing re-refined base-oils to virgin base-oil quality. There are 2 types of distillation, atmospheric and vacuum.

Atmospheric Distillation

Atmospheric distillation is generally (but not always) considered a pre-treatment step for vacuum distillation and does not require de-watered feedstock. Atmospheric distillation is carried out at normal atmospheric pressure and with temperatures up to 300°C.

Prior to the atmospheric distillation process, the feedstock can have undergone PDA treatment, but this is not an absolute pre-requisite.

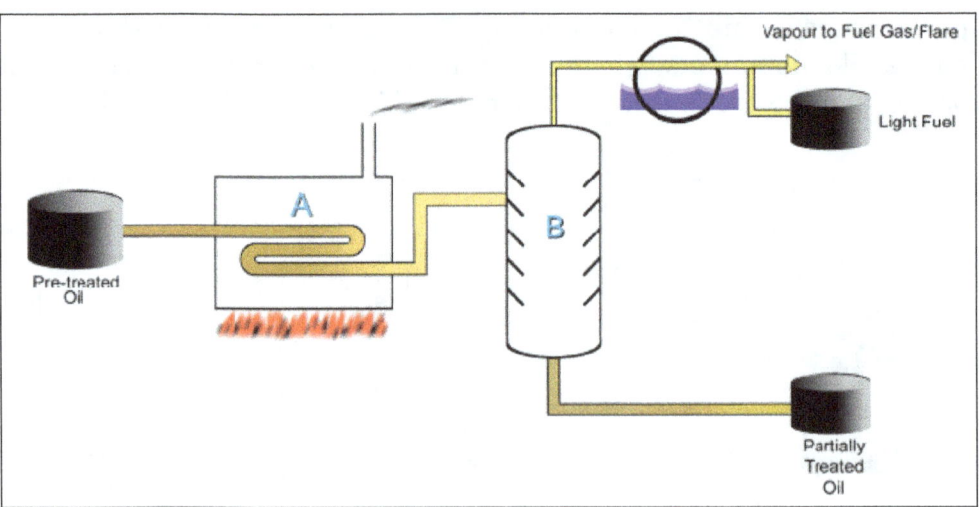

Atmospheric distillation is a relatively simple process separating lower boiling point liquids at ambient pressure. Used oil is heated (A) and charged to a distillation tower (B). Lower boiling point hydrocarbons present in the used oil (eg gases, petrol and solvents) and water are collected at the top of the tower (B). Some of these hydrocarbons can be condensed and collected for use as a fuel in the refining process.

This process is only suitable for temperatures up to 3000C, as temperatures above this can lead to "thermal cracking" of the larger molecule (higher boiling point) hydrocarbons, ie. the actual lube oil molecules we are aiming to recover.

After atmospheric distillation the oil usually undergoes vacuum distillation. Used oil can be sent directly from a "drying" process to a vacuum distillation unit without necessarily undergoing atmospheric distillation. However, it is generally accepted that water and lower boiling point hydrocarbon components be removed prior to vacuum distillation.

Vacuum Distillation

Vacuum distillation is considered the key process in used oil re-refining. If atmospheric distillation is utilised, the oil from the atmospheric distillation column is the feedstock for the vacuum distillation column. In vacuum distillation the feedstock can be separated into products of similar boiling range to better control the physical properties of the lube base stock "distillate cuts" that will be produced from the vacuum tower products.

The major properties that are controlled by vacuum distillation are viscosity, flash point and carbon residue. The viscosity of the lube-oil base-stock is determined by the viscosity of the distillate in terms of its relative viscosity separation, eg. light, medium and heavy oil.

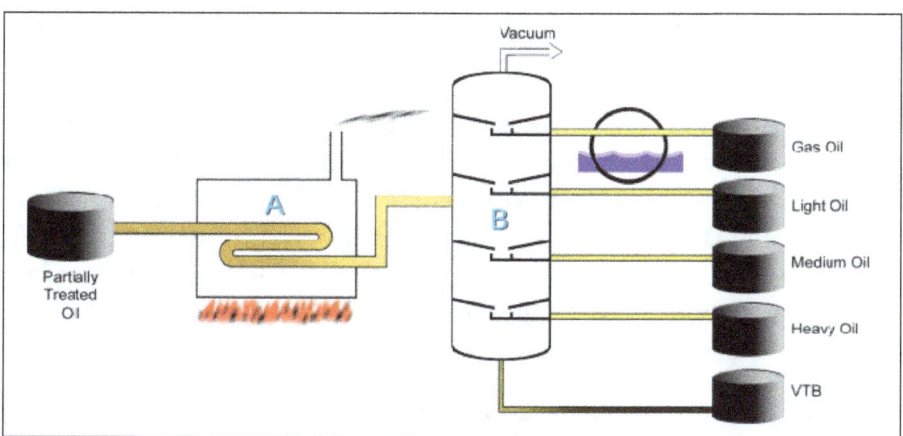

The used oil feedstock (usually from the atmospheric distillation unit) is heated in a furnace (A) and flows as a mixture of liquid and vapour to the heated vacuum distillation column (B) where the vapour portion begins to rise and the liquid falls. Steam can be added to assist vaporisation.

A vacuum is maintained in the column (2-10 mm Hg) by a vacuum system connected to the top of the tower (B). By reducing the pressure, materials normally boiling at up to about 540oC at atmospheric pressure, can be vaporised without thermal cracking.

As the hot vapours rise through the column (B), they cool and some condense to a liquid and flow back down the column. Similarly, some of the downward flowing liquids are re-vaporised by contacting the rising hot vapours. Special devices in the column allow this upward flow of vapours and downwards flow of liquids to occur continuously.

At various points in the column (B), special trays, called draw trays, are installed which permit the removal of the liquid from the column. If three cuts or "fractions" of oil are required to produce light, medium and heavy base stocks, then three draw trays are positioned appropriately. This can be reduced to two draw trays if, for example, only 2 cuts or fractions are required.

Some of the material does not boil even under this vacuum. This remains in the vacuum tower and is run out as the vacuum tower bottoms (VTBs). This material contains the heaviest molecules, including some lube oil additives and carryover contaminants not removed in the PDA process.

OIL FILTERS RECYCLING

Used oil filters are recyclable because they are made of steel, number one recycled material. They are being recycled today into new steel products, such as cans, cars, appliances and construction materials. Steel scrap is a vital ingredient in making new steel;

melting the scrap to make new steel is fundamental to energy and emissions savings and resource conservation.

While most used oil filters were thrown in the trash a short time ago, today we know they are recyclable. Some states have banned used oil filters from the landfill, while others have placed restrictions on how they can be discarded. The U.S. Environmental Protection Agency (EPA) requires used oil filters be drained of all free-flowing oil before they are discarded or recycled. Most states follow federal requirements for used oil filter disposal. Currently, U.S. manufactured oil filters are exempt from hazardous waste regulation if the oil filter is:

- Punctured through the dome end or anti-drain back valve and hot-drained.

- Hot-drained and crushed.

- Hot-drained and dismantled.

- Hot-drained using an equivalent method to remove used oil.

What is Hot-draining?

Hot-draining is defined as draining the oil filter at or near engine operating temperature, but above 60 degrees Fahrenheit. In other words, remove the filter from the engine while it is still warm, then puncture or crush and drain the filter. The EPA recommends hot-draining for a minimum of 12 hours, although specific state requirements may vary. Most of the oil is removed from the filter during hot-draining. WARNING: Use caution when hot-draining filters to avoid being burned. Protective equipment such as safety glasses and gloves should be worn to prevent injury.

Steps to Recycling Oil Filters

The three steps to recycling used oil filters are:

- Collection and transportation.

- Processing.

- Recycling by a steel mill into a new steel.

If your business changes oil commercially, it is a good idea to voluntarily collect used oil filters from Do-It-Yourselfers (DIYers). Businesses currently collecting used oil filters include auto parts stores, quick lubes and other service outlets. As state agencies see businesses voluntarily collecting filters, there will be less need for mandatory regulations.

In addition, accepting used oil filters from DIYers can be used as a tool to market your business. Studies have shown consumers are more likely to patronize businesses that offer sound environmental management practices.

Before collecting used oil filters, you should arrange for a special waste collection company to pick them up from your shop. Alternatively, if you crush them with your own on-site equipment, you can take them to a ferrous scrap processor.

Once the filters are processed, they are sent to a steel mill or foundry. Some steel mills produce flat-rolled steel products by combining scrap products and hot metal from iron ore to make products such as steel cans, cars and appliances, while others use virtually 100 percent scrap to make products such as rebar and I-beams.

About the Steel Recycling Institute

The Steel Recycling Institute (SRI), a unit of the American Iron and Steel Institute, educates the solid waste management industry, government, business and, ultimately, the consumer about the economic and environmental benefits of recycling steel. SRI works to ensure the continuing development of the steel recycling infrastructure.

APPLIANCE RECYCLING

New Orleans after Hurricane Katrina: Mounds of trashed appliances with a few smashed automobiles mixed in, waiting to be scrapped.

Appliance recycling is the process of dismantling waste home appliances and scrapping their parts for reuse. Recycling appliances for their original or other purposes, involves disassembly, removal of hazardous components and destruction of the end-of-life equipment to recover materials, generally by shredding, sorting and grading. The rate at which appliances are discarded has increased with technological advancement. This correlation directly leads to the question of appropriate disposal. The main types of appliances that are recycled are televisions, refrigerators, air conditioners, washing machines, and computers. When appliances are recycled, they can be looked upon as valuable resources. If disposed of improperly, appliances can become environmentally harmful and poison ecosystems. The strength of appliance recycling legislation and the percentage of appliances recycled varies around the world.

Disassembly

A key part of appliance recycling is the manual dismantling of each product. The disassembly removes hazardous components, while sorting out reusable parts. Procedures vary and depends on the appliance type. The amount of hazardous components able to be removed also depends on the type of appliance. Low removal rates of hazardous components reduce the recyclability of valuable materials. Each type of appliance has its own set of characteristics and components. This makes characterization of appliances essential to sorting and separating parts. Research on appliance dismantling has become an active area, intending to help recycling reach maximum efficiency.

Classification

There is a certain process used to recover materials from appliances. Parts are generally removed in order from largest to smallest. Metals are extracted first and then plastics. Materials are sorted by either size, shape, or density. Sizing is a good means of sorting to quicken future processing. It also classifies fractions that show composition. Materials report to larger or finer fractions based on original dimension, toughness, or brittleness. Shape classification contributes to the dynamics of the material. Classification by density is important when it comes to determining the use of a material.

Example:

Batteries and copper are sorted out first for quality control purposes. The materials are then compacted. Next, iron and steel (ferrous metals) are extracted using electromagnets. They are collected and made ready for sale. Then metals are separated from non-metals using eddy currents. Eddy currents are created by rapidly alternating magnetic fields, which induce metals to jump away from non-metals. Then water separation is used to sort plastics and glass from circuit boards and copper wires. Circuit boards and copper content is then sold. Plastics and glass are further compacted for reuse.

Recycling By Region

Although appliance recycling is still quite new, countries have been making the effort to enact laws and regulations regarding the electric waste. Early addressing of waste home appliance recycling started with Japan, Switzerland, Sweden, the Netherlands, and Germany.

In 2003 Waste Electrical and Electronic Equipment Directive (WEEE) passed into European Law. It sets collection, recycling and recovery targets for all types of electrical goods.

By the 1950s and 60s Japan had already become a major producer of electric appliances. The first initiatives to recycle were launched in the 70s. Due to costs, disassembly was hardly achievable. The Home Appliance Recycling Law was enacted in 1998 and came into force in 2001, and recycling of waste electrics became a legal requirement under the Specific Household Appliance Recycling Law and the Law for Promotion of Effective Utilisation Resources. Appliance manufacturers are now required to finance the recycling of their products. The Association for Electric Home Appliances is a trade group that is responsible for orphaned products.

China produces a significant share of the world's appliances. This country also has a high influx of appliance waste. There has not been much progress in appliance recycling efficiency. China's undeveloped dismantling and processing has led to elevated levels of toxins in waste appliance site vicinities. Their appliance recycling methods require severe improvement.

The United States is the largest waste appliance producer in the world, however there is still no federal law requiring appliance recycling and its legislation varies between states. On a state level, many mandatory electronic recovery programs have been implemented. There are also several commercial appliance recyclers, for example, Appliance Recycling Centers of America (ARCA). ARCA is a company based in Minneapolis, with a chain of recycling depots nationwide.

In 2003, the California Electronic Waste Recycling Act was signed. It established a new program for consumers to return, recycle, and ensure the safe and environmentally sound disposal of video display devices, such as televisions and computer monitors, that are hazardous wastes when discarded. In 2005, consumers began paying a 6-10 dollar fee when buying an electronic device. These fees are used to pay e-waste collectors and recyclers to cover their cost of managing e-waste. The EWRA classifies e-waste by dividing the products into two categories: electronic devices and covered electronic devices. Only covered electronic devices (CEDs) are included in the EWRA, however all electronic devices needed recycling measures to be taken. The CEDs include televisions and computers that have LCD displays or contain cathode ray tubes.

Australia has the same approach as the U.S. at this moment. There are several commercial appliance recyclers in Australia, as well as some organisations that remove waste appliances and offer rebates, sponsored by the government. Some retailers like Appliances Online also remove and recycle customers' old appliances using services like Sims Metal Management.

In New Zealand there is a push to keep old appliances and e waste out of landfills however this area is still largely free from legislation preventing it .Similar to Australia there are companies like Auckland based Fisher and Paykel and Appliance Recycling Ltd that remove waste appliances and recycle them using services like Phoenix Metal Recyclers.

EPR

Extended producer responsibility (EPR) is defined as an environmental protection strategy that makes the manufacturer of the appliance responsible for its entire life cycle and especially for the "take-back", recycling and final disposal of the product. Essentially, manufacturers must now finance product treatment and recycling. Countries where this strategy has been adopted for waste appliances are: Switzerland, Denmark, Netherlands, Norway, Belgium, Japan, Sweden and Germany, but it has also been expanded through legislation among certain South American countries such as Argentina, Brazil, Colombia and Peru. Countries in which EPR has long been established, demonstrate that the combination of government legislation and sound company practices can produce a higher take-back and recycling rate. An example of this is the Sony Corporation in Japan, achieving a 53% recycling rate. Other ways countries approach the issue of waste appliances is either by offering recycling facilities or banning importation. Almost all countries, at least offer facilities that aid in appliance recycling. Many implement extended producer responsibility, in addition to recycling facilities.

BATTERY RECYCLING

Battery recycling is a recycling activity that aims to reduce the number of batteries being disposed as municipal solid waste. Batteries contain a number of heavy metals and toxic chemicals and disposing of them by the same process as regular trash has raised concerns over soil contamination and water pollution.

Battery Recycling by Type

Most types of batteries can be recycled. However, some batteries are recycled more readily than others, such as lead–acid automotive batteries (nearly 90% are recycled) and button cells (because of the value and toxicity of their chemicals). Rechargeable nickel–cadmium (Ni-Cd), nickel metal hydride (Ni-MH), lithium-ion (Li-ion) and nickel–zinc (Ni-Zn), can also be recycled. There is currently no cost-neutral recycling option available for disposable alkaline batteries, though consumer disposal guidelines vary by region.

Lead–acid Batteries

These batteries include but are not limited to: car batteries, golf cart batteries, UPS batteries, industrial fork-lift batteries, motorcycle batteries, and commercial batteries. These can be regular lead–acid, sealed lead–acid, gel type, or absorbent glass mat batteries. These are recycled by grinding them, neutralizing the acid, and separating the polymers from the lead. The recovered materials are used in a variety of applications, including new batteries.

Recycling the lead from batteries.

Lead–acid batteries collected by an auto parts retailer for recycling.

The lead in a lead–acid battery can be recycled. Elemental lead is toxic and should therefore be kept out of the waste stream.

Many cities offer battery recycling services for lead–acid batteries. In some jurisdictions, including U.S. states and Canadian provinces, a refundable deposit is paid on batteries. This encourages recycling of old batteries instead of abandonment or disposal with household waste. In the United States, about 99% of lead from used batteries is reclaimed.

Businesses that sell new car batteries may also collect used batteries (or be required to do so by law) for recycling. Lead contamination of neighborhoods has resulted from the process of recycling batteries.

Silver Oxide Batteries

Used most frequently in watches, toys, and some medical devices, silver oxide batteries contain a small amount of mercury. Most jurisdictions regulate their handling and disposal to reduce the discharge of mercury into the environment. Silver oxide batteries can be recycled to recover the mercury.

Lithium Ion Batteries

Lithium-ion batteries and lithium iron phosphate ($LiFePO_4$) batteries often contain among other useful metals high-grade copper and aluminium in addition to – depending on the active material – transition metals cobalt and nickel as well as rare earths. To prevent a future shortage of cobalt, nickel, and lithium and to enable a sustainable life cycle of these technologies, recycling processes for lithium batteries are needed. These processes have to regain not only cobalt, nickel, copper, and aluminium from spent battery cells, but also a significant share of lithium. Another potentially valuable and regainable materials are graphite and manganese. Recycling processes today recover approximately 25% to 96% of the materials of a lithium-ion battery cell, depending on the separation technology. In order to achieve this goal, several steps are combined into complex process chains, especially considering the task to recover high rates of valuable materials with regard to involved safety issues.

These steps are:

- Deactivation or discharging of the battery (especially in case of batteries from electric vehicles).

- Disassembly of battery systems (especially in case of batteries from electric vehicles).

- Mechanical processes (including crushing, sorting, and sieving processes).

- Electrolyte recovery.

- Hydrometallurgical processes.

- Pyrometallurgical processes.

Specific dangers associated with lithium-ion battery recycling processes are: electrical dangers, chemical dangers, burning reactions, and their potential interactions. A complicating factor is the water sensitivity: lithium hexafluorophosphate, a possible electrolyte material, will react with water to form hydrofluoric acid; cells are often immersed in a solvent to prevent this. Once removed, the jelly rolls are separated and the materials removed by ultrasonic agitation, leaving the electrodes ready for melting down and recycling.

Pouch cells are particularly easier to recycle in this way and some people already do this to salvage the copper despite the safety issues.

As of 2019, the recycling of Li-Ion batteries in most cases does not extract lithium since lithium-ion battery technology continuously changes and processes to recycle these batteries can thus be outdated in a couple of years. Another reason why it isn't being done on a large scale is because the extraction of lithium from old batteries is 5x more expensive than mined lithium. However, it is already being done on a small scale (by some companies), an industry in expectation of large quantities of disused batteries to come.

Energy saving and effective recycling solutions for lithium-ion batteries can reduce the carbon footprint of the production of lithium-ion batteries significantly.

Battery Recycling by Location

4.5-Volt, D, C, AA, AAA, AAAA, A23, 9-Volt, CR2032, and LR44
cells are all recyclable in most countries.

In 2006, the EU passed the Battery Directive, one of the aims of which is a higher rate of battery recycling. The EU directive states that at least 25% of all the EU's used batteries must be collected by 2012, and rising to no less than 45% by 2016, of which at least 50% must be recycled.

Several sizes of button and coin cell. They are all recyclable in the UK and Ireland.

In early 2009, Guernsey took the initiative by setting up the Longue Hougue recycling facility, which, among other functions, offers a drop-off point for used batteries so they can be recycled off-island. The resulting publicity meant that a lot of people complied with the request to dispose of batteries responsibly.

From April 2005 to March 2008, the UK non-governmental body WRAP conducted trials of battery recycling methods around the UK. The methods tested were: Kerbside, retail drop-off, community drop-off, postal, and hospital and fire station trials.

The kerbside trials collected the most battery mass, and were the most well-received and understood by the public. The community drop-off containers which were spread around local community areas were also relatively successful in terms of mass of batteries collected. The lowest performing were the hospital and fire service trials (although these served their purpose very well for specialized battery types like hearing aid and smoke alarm batteries). Retail drop off trials were the second most effective (by volume) method but one of the least well received and used by the public. Both the kerbside and postal trials received the highest awareness and community support.

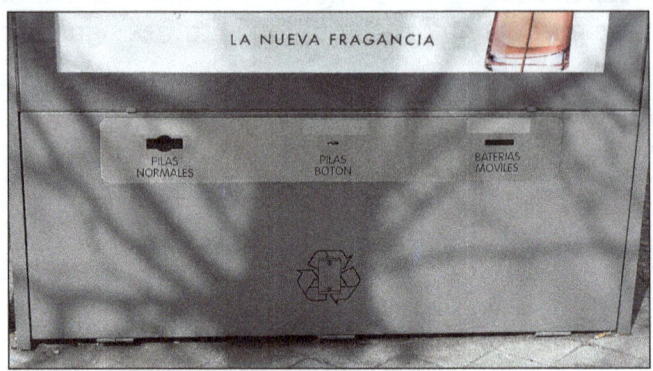

A battery recycling station at a bus stop in Madrid.

Household batteries can be recycled in United Kingdom at council recycling sites as well as at some shops and shopping centres—e.g., Dixons, Currys, The Link and PC World.

A scheme started in 2008 by a large retail company allowed household batteries to be posted free of charge in envelopes available at their shops. This scheme was cancelled at the request of the Royal Mail because of hazardous industrial battery waste being sent as well as household batteries.

An EU directive on batteries that came into force in 2009 means producers must pay for the collection, treatment, and recycling of batteries. This has yet to be ratified into UK law however, so there is currently no real incentive for producers to provide the necessary services.

From 1 February 2010, batteries can be recycled anywhere the "Be Positive" sign appears. Shops and online retailers that sell more than 32 kilograms of batteries a year must offer facilities to recycle batteries. This is equivalent to one pack of four AA batteries a day. Shops which sell this amount must by law provide recycling facilities as of 1 February 2010.

In Great Britain an increasing number of shops (Argos, Homebase, B&Q, Tesco, and Sainsbury's) are providing battery return boxes and cylinders for their customers.

The rechargeable battery industry has formed the Rechargeable Battery Recycling Corporation (RBRC), which operates a battery recycling program called Call2Recycle throughout the United States and Canada. RBRC provides businesses with prepaid shipping containers for rechargeable batteries of all types while consumers can drop off batteries at numerous participating collection centers. It claims that no component of any recycled battery eventually reaches a landfill. Other programs, such as the Big Green Box program, offer a recycling option for all chemistries, including primary batteries such as alkaline and primary lithium.

A study estimated battery recycling rates in Canada based on RBRC data. In 2002, it wrote, the collection rate was 3.2%. This implies that 3.2% of rechargeable batteries were recycled, and the rest were thrown in the trash. By 2005, it concluded, the collection rate had risen to 5.6%.

In 2009, Kelleher Environmental updated the study. The update estimates the following. "Collection rate values for the 5 [and] 15-year hoarding assumptions respectively are: 8% to 9% for NiCd batteries; 7% to 8% for NiMH batteries; and 45% to 72% for lithium ion and lithium polymer batteries combined. Collection rates through the [RBRC] program for all end of life small sealed lead acid (SLA) consumer batteries were estimated at 10% for 5-year and 15-year hoarding assumptions. It should also be stressed that these figures do not take collection of secondary consumer batteries through other sources into account, and actual collection rates are likely higher than these values."

Batteries collected in the United States are increasingly being transported to Mexico for recycling as a result of a widening gap between the strictness of environmental and labor regulations between the two countries.

In 2015, Energizer announced availability of disposable AAA and AA alkaline batteries made with 3.8% to 4% (by weight) of recycled batteries, branded as EcoAdvanced.

Japan does not have a single national battery recycling law, so the advice given is to follow local and regional statutes and codes in disposing batteries. The Battery Association of Japan (BAJ) recommends that alkaline, zinc-carbon, and lithium primary batteries can be disposed of as normal household waste. The BAJ's stance on button cell and secondary batteries is toward recycling and increasing national standardisation of procedures for dealing with these types of batteries.

In April 2004, the Japan Portable Rechargeable Battery Recycling Center (JBRC) was created to handle and promote battery recycling throughout Japan. They provide battery recycling containers to shops and other collection points.

The environmental pollution caused by the valuable chemical components such as cobalt, copper, lithium, mixture of organic electrolyte and salts of either low quality or spent lithium-ion batteries (LiBs) deposited into the environments necessitates responsive recovery technologies. Since Sony made the first commercial lithium-ion cell in 1991, it has been accorded more attention being superior to other types of batteries in terms of energy density, which is a critical parameter for portable electronics as well as hybrid and electric vehicles. Lithium ion batteries are the systems preferred as electrochemical power sources in portable batteries segment such as mobile tele-phones, personal computers, video-cameras and other modern-life appliances as well as in vehicles with electric drive due to its favorable characteristics. As LiBs progres-sively dominate, the amounts of valuable chemical components that will be deposited will be proportional to the number of LiBs used after their life-span has expired. There-fore, recycling that constitutes the most generally acceptable environmentally friendly method of managing these wastes must be taken serious, to minimize environmental toxicity, for economic gains and reduction in dependence on foreign resources or on virgin materials for productions in the industry as well as for sustainability of the nat-ural resources. The methods could be on the laboratory scale, industrial or commercial scale level. These as-recovered metals or their respective compounds (cobalt, lithium, manganese, and nickel) are not only valuable metals but are alternative precursors for new batteries formulations. Thus, several attempts have been made to review the old processes considered green and non-green chemistries to either improve on the exist-ing ones or propose new recovery processes that are considered simple and of industri-al-scale. However, the cells used in cell phones and laptops are not fully recycled and consequently causing unsustainable open loop in the industrial cycle.

Spent LiBs have been classified as non-environmentally hazardous wastes or rather call "green batteries" and thus safe for disposal in the normal municipal waste stream unlike other battery chemistries that contain Cd, Pb or Hg, the presence of flammable and toxic elements or compounds may make their safe disposal to become a serious problem. For instance, the mixture of dimethyl carbonate (DMC) and ethylene carbon-ate (EC) used as solvent is flammable, while the polyvinylidene fluoride (PVDF) used as binder irrespective of its percentage in the battery formulation is toxic when burns consequent to the release of gaseous HF. Besides, the NMP commonly used as a solvent for the electrode active materials (cathode and anode) fabrication during slurry prepa-ration has been reported as toxic and therefore environmentally incompatible. As there is a general saying and belief that "health is wealth", similarly, "healthy environment is a wealthy environment". On the other hand, "polluted environment is an unhealthy and un-wealthy environment". Therefore, recycling is of great importance to save our immediate environment and for waste management sustainability.

Structural Composition of Lithium-ion Battery

All batteries consist of cathode, anode, electrolyte mixture and separator. The cath-ode has the aluminium foils coated with a mixture of the active material, $LiCoO_2$ or $LiNi_{1/3}Mn_{1/3}Co_{1/3}O_2$ depending on the type, PVdF or PTFE, carbon graphite, while the anode is a copper foil coated with blended slurry of carbon graphite and PVdF or PTFE.

The electrolyte mixture consists of the water-proned electrolyte salt, $LiPF_6$ and or-ganic solvents dissolved in varying ratios such as 1:1:1 (v/v) for 1M LiPF6, dimethyl carbonate (DMC) and ethylene carbonate (EC) respectively. In addition, other lithium salts used for lithium-ion battery are $LiAsF_6$, $LiClO_4$, and $LiBF_4$, while the organic solvents among others are propylenecarbonate with dimethoxyethane (PC–DME), γ-butyrolactone with tetrahydrofuran (BL–THF) and dioxolane (1, 3-D) according to Contestabile et al.

The separator is a non-conductor that separates the two electrodes from each other. The structure of a cylindrical lithium-ion battery according to Nishi is represented in figure.

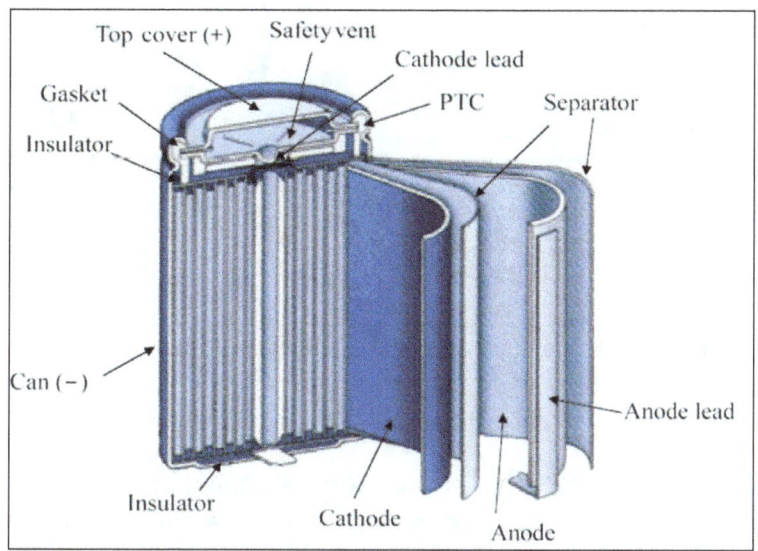

The structure of a cylindrical lithium-ion battery.

Chemical Reaction of a Typical Lithium-ion Battery

Lithium ions move from the negative electrode to the positive electrode during the discharge process through the nonaqueous electrolyte and separator diaphragm and then undergo reversible reaction when charging. The ionic chemical reactions are shown in equations.

The cathodic half reaction:

$$LiMO_2 \underset{Charge}{\overset{Discharge}{\rightleftarrows}} Li_{1-x}MO_2 + xLi^+ + xe^-.$$

The anodic half reaction:

$$xLi^+ + xe^- + 6C \underset{Charge}{\overset{Discharge}{\rightleftarrows}} Li_xC_6.$$

The overall reaction:

$$LiMO_2 + 6C \underset{Charge}{\overset{Discharge}{\rightleftarrows}} Li_{1-x}MO_2 + Li_xC_6.$$

Where M reprersents Mn, Ni or Co depending on the cathode active material.

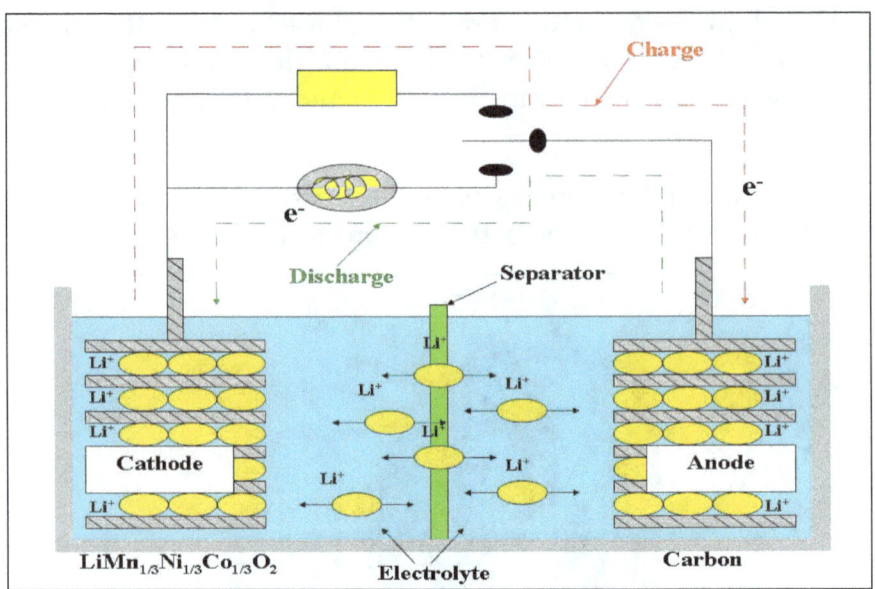

Schematic diagram of the chemical reaction of a lithium-ion battery.

Processes for Recovery of Lithium Ion Batteries

Recycling technologies, irrespective of the processes must amongst others achieve the reduction in the volume of the scraps or cases, selective separation of the valuable components. The physical and chemical processes are generally the two categories of processes employed in the laboratory and industry to recycle all kinds of batteries.

Physical Processes

The physical processes are generally dissolution, manual or mechanical separation and pyrolysis. For instance, Contestabile et al and Bankole and Lei extracted the electrolyte solution into organic solvents such as ethanol or iso-butylalcohol/water after manually or mechanical dismantling LiBs and this enhanced reduction in the environmental pollution caused by the hydrolysis of electrolyte salt, $LiPF_6$ and also the toxic electrolyte mixture. Interestingly, innovative conversion of $LiPF_6$ to useful compound such as Li_2SiF_6 was achieved for the first time.

Hydrometallurgical Process

In hydrometallurgical method, mechanical separation was employed as pretreatment by subjecting LiBs to skinning, crushing removing of crust, sieving and separation of both anode and cathode material for easy recovery of the valuable components of the batteries. However, safety precautions are required due to flammability of the electrolyte mixture. Although the stress in manual separation will be reduced, the components of the batteries may not be fully separated from one another due to the structural arrangement of the LiBs.

Dissolution Process

This process recently dominates and enhances effectiveness with maximum recovery of valuable components from batteries. The adhesive force from the PVdF holding the electrode active materials (anode and cathode) unto the current collectors is weakened. Therefore, the choice of suitable organic solvents capable of dissolving the binder, PVdF or PTFE becomes very important during recovery processes. Among these suitable solvents that have already been tested and found effective are N, N-dimethylformamide (DMF), N, N-dimethyl acetamide (DMAC), N-methylpyrrolidone (NMP) and dimethylsulfoxide (DMSO) with their order of effectiveness in dissolving the adhesive investigated. For instance, $LiCoO_2$ was recovered from LiBs with the solubility of PVdF in the first three solvents recorded as DMAC > DMF > NMP. N-methylpyrrolidone separated both $LiMn_{1/3}Ni_{1/3}Co_{1/3}O_2$ and $LiCoO_2$, from LiBs at 40 °C for 15 minutes and at 100 °C for 1 h, respectively. Although the powders were effectively recovered, the cost of buying 1L of NMP which is about $ 207.90 makes its application not cost-effective and suitable for a large scale recovery operation. Among all these solvents, DMSO used at 60 °C for 85 minutes could be the most suitable for its cheapness ($ 144.54/ L), non-toxicity and environmental safety. Moreover, the clean and shiny current collectors (Aluminium foils) obtained after the separation could be used for other applications in the laboratory and industries. The flow-sheet for the recycling of LiBs by dissolution method is shown in figure.

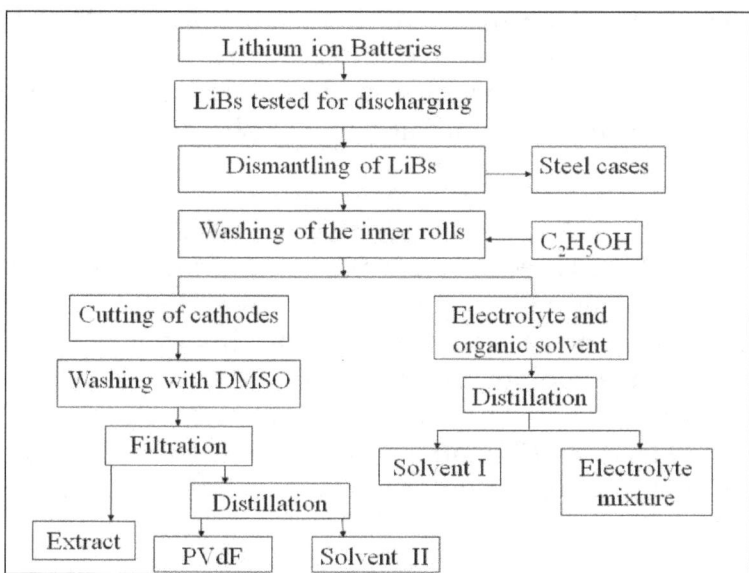

Flow sheet for the recovery of valuable components from LiBs by dissolution process.

Pyrolysis or Pyrometallurgical Process

The name comes from the two words "pyro" and "lysis" meaning "fire" and "decomposition", respectively. Therefore, this process decomposes the components of the LIBs by heating to high temperatures under heat and pressure. Pyrometallurgical process has been associated with high air emission of dioxins, chloride compounds and mercury, and therefore requires strict standard for air filtration systems to avoid pollution. It was used as pre-treatment for waste batteries before leaching process, especially to remove Hg, papers and plastics under a controlled atmosphere.

Chemical Processes

The chemical processes are mainly hydrometallurgical methods involving acid or base leaching, solvent extraction, chemical precipitation, bioprocess and electrochemical process or combination of the processes. The multiple-steps will consume more chemicals.

Hydrometallurgical Processes

The scraps of the spent LiBs were put in either acid or alkaline solution to dissolve the metallic fraction of the batteries to recover valuable components. Hydrometallurgical was used on the basis of its simplicity, environmentally friendly due to waste water and air emission minimization, adequate recovery of valuable metals with high purity and low energy requirements. For instance, cobalt-containing slag was treated through hydrometallurgical process by Lain and Espinosa et al.

This process also used the mixture of H_2SO_4 and H_2O_2 to recover Li and Co from LiBs and achieved full recovery of the metals within 10 min at 75 °C with an agitation of 300 rpm. However, the thermal pretreatment of $LiCoO_2$ particles to remove carbon and organic binder before chemical leaching significantly reduced the leaching efficiency. Also, $LiPF_6$ decomposed into lithium fluoride and phosphorus pentafluoride during crushing process.

Also, with an enhanced leaching efficiency, mixture of an easily degradable organic acid DL-malic acid and H_2O_2 was used to recover Co and Li from LiBs. Instead of DL-malic acid with H_2O_2, both Co and Li were effectively recovered using citric acid and H_2O_2. Kang et al leached cobalt-containing powder from LiBs with H_2SO_4 and H_2O_2 to recover cobalt sulfate, while addition of oxalic acid to the filtrate from another powder produced crystalline cobalt oxalate, which was then heated to produce Co_3O_4. Zhang et al recovered Co and Li using HCl solution. The Co in the leached liquor was selectively extracted with PC-88A in kerosene and then as cobalt sulfate with high purity, while Li was obtained as $LiCO_3$.

A combination of ultrasonic washing, acid leaching and precipitation was proposed by Li et al to recover Co from spent LIBs. The ultrasonic washing improved the recovery efficiency of Co, reduced energy consumption as well as environmental pollution.

This process was considered feasible for recycling spent LIBs for scale-up operation. A recycling process that combined hydrometallurgical and sol–gel steps in addition to other general steps was also used to recover Co and Li from LiBs. The acid media (hydrogen peroxide in HNO_3) used enhanced the leaching efficiency. A gelatinous precursor was prepared by adding citric acid to the leaching liquor to obtain amorphous citrate precursor process (ACP), followed by pyrolysis to obtain pure crystalline $LiCoO_2$.

Combined Acid-alkaline with Organic Solvents Process

As a means of advancing the process of recycling spent lithium-ion batteries, combined acids-alkaline and organic solvents was used for safety, simplicity and other benefits observed in other methods. Lithium, Ni, Mn and Co were leached from $LiMnNiCoO_2$ using HNO_3 and then precipitant, NaOH by Castillo et al, while Consestabile et al also leached $LiCoO_2$ with HCl and then precipitated the cations with NaOH solution. In similar steps, the batteries inner rolls were treated with NaOH to dissolve the aluminium foil to separate the cathode material powders from other components. The powder obtained was burnt to remove PVdF, followed by dissolution to produce sulfate solution. Cobalt in the solution was deposited as oxalate, while Acorga M5640 and Cyanex272 (di-(2,4,4 trimethyl pentyl) phosphoric acid) were used to selectively extract small quantities of Cu^{2+}, Co^{2+} and Ni^{2+} ions in the solution. Wang et al selectively used KMnO4 to recover Mn as MnO2 and manganese hydroxide from the leaching liquor, while dimethylglyoxime was used to recover Ni. Cobalt was precipitated as cobalt hydroxide, while addition of a saturated Na_2CO_3 solution to the liquor precipitated Li as Li_2CO_3. The process can be represented by the flow-sheet in figure.

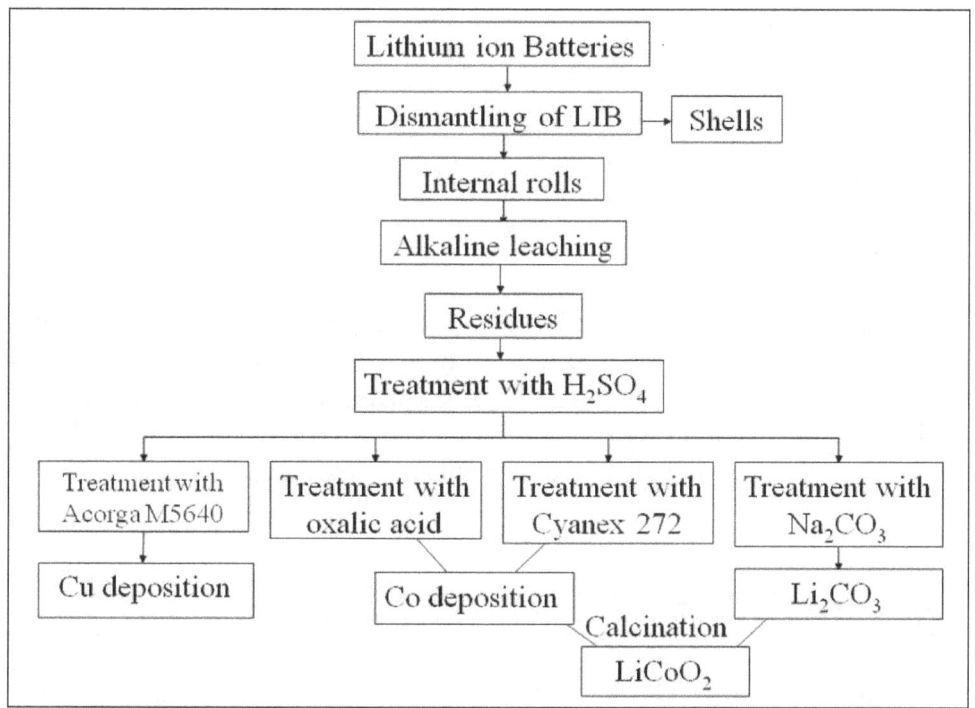

Flow sheet for general acid-alkaline with selective recovery and recycling of LiBs.

Bio-Metallurgical Process

Compared with the aforementioned pyrometallurgical, hydrometallurgical processes, bioprocess was considered as having higher efficiency, low cost and environmentally compatible. The process used bacteria and inorganic chemical solutions. For example, acidithiobacillus ferrooxidans utilized elemental sulfur and ferrous ion to produce metabolites, H_2SO_4 and ferric ion in the leaching medium to recover Li and Co from $LiCoO_2$ of LiBs. The metabolites enhanced the dissolution of metals from the batteries. Comparatively, bio-dissolution of Co was faster than Li. Xin et al also recovered Co and Li from the spent LiBs through the same processes. However effective the procedure may be, the cost of culturing the enzymes or bacteria may somehow hinder its commercial operation.

Electrochemical Process

Electrochemical methods have been used to recover metals from the leached liquor of the cathode active materials of LiBs. Meanwhile, it was impossible to recover Ni directly by the method from the liquor obtained. Therefore, Ni was first separated from Co by solvent extraction, followed by its recovery through galvanostatic and potentiostatic electrowinning. Also, Freitas and Garcia electrochemically recovered Co, while combination of the electrochemical and hydrothermal methods were used to recover both Co and Ni from $LiCoO_2$ and $(LiCo_xNi_{(1-x)})O_2$ in the Li ion and Li polymer batteries, respectively. The ionic equations for the electrochemical reactions of a divalent cation during electrolytic recycling process could follow:

$$Anode: M^{2+} + 2H_2O \rightarrow MO_2 + 4H^+ + 2e^-$$

$$Cathode: 2H^+ + 2e^- \rightarrow H_2$$

$$Overall: M^{2+} + 2H_2O \rightarrow MO_2 + 2H^+ + H_2$$

Pyrometallurgical Process

The process chemically recovered valuable components of the waste materials or concentrates at elevated temperatures. Pauline et al fused the mixture of active mass (cathode and anode) and electrolyte with $KHSO_4$ in a furnace. Although precaution was taking to avoid reduction of sulfate to SO_2 of sulfide, industrial dumps like CaF_2, $Ca_3(PO_4)_2$ and other byproducts were generated along with the desired products.

FLUORESCENT LAMP RECYCLING

Fluorescent lamp recycling is the recovery of the materials of a spent fluorescent lamp for the manufacture of new products.

Glass tubing can be turned into new glass articles, brass and aluminum in end caps can be reused, the internal coating can be reprocessed for use in paint pigments, and the mercury contained in the lamp can be reclaimed and used in new lamps. In the United States, about 620 million fluorescent lamps are discarded annually; proper recycling of a lamp prevents emission of mercury into the environment, and is required by most states for commercial facilities. The primary advantage of recycling is diversion of mercury from landfill sites, but the actual scrap value of the materials salvaged from a discarded lamp is often insufficient to offset the cost of recycling.

Mercury in Lamps

The amount of mercury in a fluorescent lamp varies from 3 to 46 mg, depending on lamp size and age. Newer lamps contain less mercury and the 3–4 mg versions are sold as low-mercury types. A typical 2006-era 4 ft (122 cm) T-12 fluorescent lamp (i.e. $F_{34}T_{12}$) contains about 5 milligrams of mercury. In early 2007, the National Electrical Manufacturers Association in the US announced that "Under the voluntary commitment, effective April 15, 2007, participating manufacturers will cap the total mercury content in CFLs under 25 watts at 5 milligrams (mg) per unit. CFLs that use 25 to 40 watts of electricity will have total mercury content capped at 6 mg per unit."

Only a few tenths of a milligram of mercury are required to maintain the vapor, but lamps must include more mercury to compensate for the part of mercury absorbed by internal parts of the lamp and no longer available to maintain the arc. Manufacturing processes have been improved to reduce the handling of liquid mercury during manufacture and improve accuracy of mercury dosing.

Mercury-free discharge lamps have considerably lower production of visible light, reduced to about half; thus, mercury remains an essential component of efficient fluorescent lamps.

Broken Lamps

A broken fluorescent tube will release its mercury content. Safe cleanup of broken fluorescent bulbs differs from cleanup of conventional broken glass or incandescent bulbs, avoiding the use of vacuum cleaners, in favour of sticky tape to recover small particles, and ensuring that fans and air conditioning are turned off. Approximately 99% of the mercury is typically contained in the phosphor, especially on lamps that are near their end of life.

Phosphors

Lamps made up to the 1940s used toxic beryllium compounds, which were implicated in the deaths of factory workers. Today it is very unlikely that one would encounter any such lamps.

Other toxic elements such as arsenic, cadmium, and thallium were formerly used in phosphor manufacture. Modern halophosphate phosphors resemble the chemistry of tooth enamel. The rare-earth doped phosphors are not known to be harmful.

Mercury Containment

When a modern fluorescent tube is discarded, the main concern is the mercury, which is a significant toxic pollutant. One way to avoid releasing mercury into the environment is to combine it with sulfur to form mercury sulfide, which will prevent vapor release and is insoluble in water. One advantage of sulfur is its low cost. The reaction is shown with the equation:

$$Hg + S \rightarrow HgS.$$

The easiest way to combine sulfur and mercury is to cover a group of fluorescent tubes with sulfur dust (sometimes called "flowers of sulfur") and to break the tubes; when the glass fragments are put into a bag to continue with the reaction, the mercury will combine with sulfur without any other action. The glass can be recycled where an appropriate facility exists. A quantity of 25 kilograms (55 lb) of dust sulfur is enough for 1000 tubes.

Disposal Methods

The disposal of phosphor and mercury toxins from spent tubes can be an environmental hazard. Governmental regulations in many areas require special disposal of fluorescent lamps separate from general and household wastes. For large commercial or industrial users of fluorescent lights, recycling services are available in many nations, and may be required by regulation. In some areas, recycling is also available to consumers.

Spent fluorescent lamps are typically packaged prior to transport to a recycling facility in one of three ways: Boxed for bulk pickup, by using a prepaid lamp recycling box, or crushed onsite before pickup. A fluorescent lamp crusher can attach directly to a disposal drum and isolate the dust and mercury vapor. In some states, drum-top crushers and end-user crushing of lamps are not allowed. Minnesota Department of Health Drum Top Bulb Crusher Demonstration.

TEXTILE RECYCLING

Textile recycling is the process of recovering fiber, yarn or fabric and reprocessing the textile material into useful products. Textile waste products are gathered from different sources and are then sorted and processed depending on their condition, composition, and resale value. The end result of this processing can vary, from the production of energy and chemicals to new articles of clothing.

Textiles collection boxes in Brussels.

Due to a recent trend of over consumption and waste generation in global fashion culture, textile recycling has become a key focus of worldwide sustainability efforts. Globalization has led to a "fast fashion" trend where clothes are considered by many consumers to be disposable due to their increasingly lower prices. The development of recycled technology has allowed the textile industry to produce vast amounts of products that deplete natural resources. Textile recycling techniques have been developed to cope with this increase of textile waste and new solutions are still being researched. Recently, certain clothing retailers have embraced this recycling effort and now publicly advertise products that are made of recycled textile material in accordance with shifting consumer expectations.

Starting Material

Most materials used in textile recycling can be split into two categories: Pre-consumer and post-consumer waste.

Pre-consumer Waste

Pre-consumer or post-industrial waste consists of textile waste produced at the industrial stage of the production of textile material. Typically, these byproducts are produced by the textile, garment, cotton, and fiber industries and are repurposed by the furniture, home building, automotive, and other industries.

Post-consumer Waste

Post-consumer or waste consists of discarded garments or household articles made from manufactured textiles. These unwanted articles are typically worn out or damaged. Some post-consumer waste is directed towards second hand retailers to be sold again. Some of this waste is collected in municipal collection bins, but the majority of this waste is found in landfills.

The clothing brand The North Face introduced a program called "Clothes the Loop" in 2013 that allows consumers to recycle post-consumer waste from any brand at any of their retail locations across the United States. This mirrors similar services by charity organizations such as Goodwill Industries and The Salvation Army in the United States. Across the globe, charitable organizations and businesses such as thrift stores have created specially marked collection bins that allow the public to dispose of post-consumer waste so that it can be reused and repurposed.

Processing

Sorting

When recycling post-consumer textile waste, the sorting process is represented as a pyramid model in terms of the volume of material. At the base of the pyramid - and largest volume - is crude sorting, followed by exportation of second-hand clothing, conversion to new products, wiping and polishing cloths, landfill incineration for energy, and lastly diamonds. Typically within the pyramid model it is found that the volume of clothing items is inversely proportional to its monetary value, moreover meaning that despite diamonds making up the smallest sector (1-2%) of the sorting process they tend to be the most profitable.

Crude Sorting

Within crude sorting, waste items are often manually separated into distinct categories whilst also removing bulkier items, such as coats and blankets. The categories of textile waste may be divided based upon elements such as material, condition, quality, or clothing item such as shirts. Employees with the most expertise perform the most detail-oriented distinctions such as being able to distinguish cashmere from wool by touch. Along the crude sorting process, recycled textiles are also assigned categoric grades representing their commercial value based upon various fiber characteristics such as length, color, and the homogeneity of its chemical composition.

Second Hand Clothing Exportation Markets

The exportation of second-hand clothing is a growing global market; the trade market value doubled between the years 2007 and 2012 based upon declared reports alone. The exportation trend is most commonly from Western countries to developing countries or those experiencing disaster relief, with the United States of America being responsible for 45% of the total volume of Western exportation. In Africa specifically, Western clothing is a high commodity that imports $61.7 million of sales annually and in Sub-Saharan Africa these exports account for over a third of the total purchased garments.

Textile Conversion to New Products

Clothing items that are not able to be resold may be converted into new products using their recycled fibers. Shoddy and mungo are the two main results of this process.

Shoddy is one of the most historical examples of textile recycling and refers to creating yarn products from the old materials. Mungo was invented after shoddy and refers to the process of using clippings of textiles when making wool, which is mostly exported to European countries due to the need for wool in the cooler temperatures and flammability regulations. Specific examples of products being produced using either shoddy or mungo include luxury blankets in Italy, fibers within US dollars, and the phenomena of sustainable fashion trends.

Wiping and Polishing Cloths

Textiles that are deemed to be un-wearable during the sorting phase may be used to create wiping and polishing cloths that are called snakes, made from oleophilic and hydrophilic fibers. T-shirts are largely used when creating these cloths due to the naturally absorbent cotton fibers that have proven useful for cleaning up small oil spills within the oil industry.

Creating Energy from Landfill Incineration

The textile materials that are not found to have a viable market in any of the above categories are either sent to the landfill or are incinerated to produce electrical energy. Though incinerating municipal solid waste (MSW) is not yet feasible in the United States, it has been prolific in countries such as Denmark, Japan, and Switzerland where over two thirds of MSW is incinerated. In terms of calories, the energy values of burning MSW have been comparable with oil but there are still obstacles with increasing incineration efficiency and reducing harmful byproducts of the incineration such as noxious gases and ash.

Diamonds

Diamonds make up the smallest sector of the sorting process, constituting the rare items of clothing that are considered collectibles and can be resold with a high profit margin. Diamonds are usually hunted and sold by small family businesses, the majority of these lie in Japan where consumers value Western brands such as Ralph Loren, Levi's. and Harley Davidson. Whereas, the smaller diamond markets in America place a higher value and priority on items such as Italian leather and French Couture.

Methods

Mechanical

Mechanical processing is the most commonly used technique to recycle textiles. Companies in the United States used about 7.6 million bales of cotton to manufacture textiles with each bale weighing 500 pounds. The cotton can be recycled through mechanical means after separating it from different materials. However, some plants can still process recycled material that is not purely made of cotton such as 98% cotton and 2% spandex. After an initial sorting, the raw material is further sorted by color to avoid re-dying and bleaching. Once done, the textile material is shredded and separated into fibers. The end product at this point is not usable yet and needs to be aligned before spinning. This process is known as carding. Now, the fibers are spun along with some virgin cotton fibers since recycled cotton fibers are shorter and lower in quality. Another commonly used material in mechanical processing is polyester. With this process, the recycled materials are not polyester textiles, but plastic bottles. Both are made of the same material known as polyethylene terephthalate (PET). Once the materials are sent to the facility, first they are sorted by color and type. Similar to cotton, the PET plastic is shredded into slices and washed to remove contaminants. The dried, shredded plastic is molded into PET pellets and undergoes extrusion to create new fibers.

Chemical

Chemical processing is typically used on synthetic fibers such as Polyethylene terephthalate (PET) as these fibers can undergo a breaking down and recreation process. This process is not yet widely used, but there are companies that are researching and implementing chemical recycling. The major small scale production sites are from Eco Circle, Worn Again, Evrnu, and Ioncell. In the case of PET, the starting materials are first broken down to monomers. This is done by using chemicals that facilitate glycolysis, methanolysis, hydrolysis, and ammonolysis. This act of depolymerization also removes contaminants from the starting material such as dyes and unwanted fibers. From here, the material is polymerized to be used to produce textile products. Unlike the mechanical method of recycling, chemical recycling produces high quality fibers similar to the virgin fiber used. Therefore, no new fibers are needed to support the product of the chemical process. Different chemicals and pathways are used for other materials such as nylon and cellulose based fibers, but the overall structure of the process is the same.

Applications of Recycled Textiles

Many clothing retailers across the world have implemented recycled textiles in their products. Companies like These retailers cater to environmentally conscious consumers by producing Sustainable Fashion.

Products

Companies such as Patagonia, Everlane, H&M, Lindex, Pure Waste, and Heavy Eco sell sustainable clothes. These companies incorporate materials derived from textile post-consumer waste as well as recycled plastics into the clothes they sell.

Within Scandinavia specifically there are prolific applications of recycled textiles that have created mainstream market products. In Sweden, companies such as H&M and

Lindex are including pre and post-consumer waste fibers within their new clothing lines. Similarly in Finland, Pure Waste is a clothing enterprise that creates t-shirts from recycled fibers in their 95% wind powered factories. Aside from clothing, Egetæpper is a Danish carpet manufacturing company that makes their carpets using the fibers from recycled fishnets.

TIRE RECYCLING

Tyre Arm Chair.

Tires are among most problematic sources of waste. Progress in recycling has resulted in a major reduction in dumping.

Tire recycling, or rubber recycling, is the process of recycling waste tires that are no longer suitable for use on vehicles due to wear or irreparable damage. These tires are a challenging source of waste, due to the large volume produced, the durability of the tires, and the components in the tire that are ecologically problematic.

Because tires are highly durable and non-biodegradable, they can consume valued space in landfills. In 1990, it was estimated that over 1 billion scrap tires were in stockpiles in the United States. As of 2015, only 67 million tires remain in stockpiles. From 1994 to 2010, the European Union increased the amount of tires recycled from 25% of annual discards to nearly 95%, with roughly half of the end-of-life tires used for energy, mostly in cement manufacturing.

Newer technology, such as pyrolysis and devulcanization, has made tires suitable targets for recycling despite their bulk and resilience. Aside from use as fuel, the main end use for tires remains ground rubber.

In 2017, 13% of U.S. tires removed from their primary use were sold in the used tire market. Of the tires that were scrapped, 43% were burnt as tire-derived fuel, with cement manufacturing the largest user, another 25% were used to make ground rubber, 8% were used in civil engineering projects, 17% were disposed of in landfills and 8% had other uses.

Tire Life Cycle

The tire life cycle can be identified by the following six steps:

- Product developments and innovations such as improved compounds and camber tire shaping increase tire life, increments of replacement, consumer safety, and reduce tire waste.

- Proper manufacturing and quality of delivery reduces waste at production.

- Direct distribution through retailers, reduces inventory time and ensures that the life span and the safety of the products are explained to customers.

- Consumers' use and maintenance choices like tire rotation and alignment affect tire wear and safety of operation.

- Manufacturers and retailers set policies on return, retread, and replacement to reduce the waste generated from tires and assume responsibility for taking the 'tire to its grave' or to its reincarnation.

- Recycling tires by developing strategies that combust or process waste into new products, creates viable businesses, and fulfilling public policies.

Landfill Disposal

Tires are not desired at landfills, due to their large volumes and 75% void space. Tires can trap methane gases, causing them to become buoyant, or bubble to the surface. This 'bubbling' effect can damage landfill liners that have been installed to help keep landfill contaminants from polluting local surface and ground water.

Shredded tires are now being used in landfills, replacing other construction materials, for a lightweight back-fill in gas venting systems, leachate collection systems, and operational liners. Shredded tire material may also be used to cap, close, or daily cover landfill sites. Scrap tires as a back-fill and cover material are also more cost-effective, since tires can be shredded on-site instead of hauling in other fill materials.

Stockpiles and Legal Dumping

Tire stockpiles create a great health and safety risk. Tire fires can occur easily, burning for months and creating substantial pollution in the air and ground. Recycling helps to reduce the number of tires in storage. An additional health risk, tire piles provide harborage for vermin and a breeding ground for mosquitoes that may carry diseases. Illegal dumping of scrap tires pollutes ravines, woods, deserts, and empty lots; which has led many states to pass scrap tire regulations requiring proper management. Tire amnesty day events, in which community members can deposit a limited number of waste tires free of charge, can be funded by state scrap tire programs, helping decrease illegal dumping and improper storage of scrap tires.

Used tires in foreground waiting to be shredded and shredded tires in background.

Tire storage and recycling are sometimes linked with illegal activities and lack of environmental awareness.

Uses

Although tires are usually burnt, not recycled, efforts are continuing to find value. Tires can be reclaimed into, among other things, the hot melt asphalt, typically as crumb rubber modifier—recycled asphalt pavement (CRM—RAP), and as an aggregate in portland cement concrete Efforts have been made to use recycled tires as raw material for new tires, but such tires may integrate recycled materials no more than 5% by weight, and tires that contain recycled material are inferior to new tires, suffering from reduced tread life and lower traction. Tires have also been cut up and used in garden beds as bark mulch to hold in the water and to prevent weeds from growing. Some "green" buildings, both private and public, have been made from old tires.

Pyrolysis can be used to reprocess the tires into fuel gas, oils, solid residue (char), and low-grade carbon black, which cannot be used in tire manufacture. A pyrolysis method which produces activated carbon and high-grade carbon black has been suggested.

Cement Manufacturing

Old tires can be used as an alternative fuel in the manufacturing of Portland cement, a key ingredient in concrete. Whole tires are commonly introduced into cement kilns, by rolling them into the upper end of a preheater kiln, or by dropping them through a slot midway along a long wet kiln. In either case, the high gas temperatures (1000–1200 °C) cause almost instantaneous, complete and smokeless combustion of the tire. Alternatively, tires are chopped into 5–10 mm chips, in which form they can be injected into a precalciner combustion chamber. Some iron input is required in manufacturing cement, so the iron content of steel-belted tires is beneficial to the process.

Used tires being fed mid-kiln to a pair of long cement kilns.

Tire-derived Products

Shredded tires.

Tires can be reused in many ways, although most used tires are burnt for their fuel value. it is stated that markets ("both recycling and beneficial use") existed for 80.4% of scrap tires, about 233 million tires per year. Assuming 22.5 pounds (10.2 kg) per tire, the 2003 report predicts a total weight of about 2.62 million tonnes (2,580,000 long tons; 2,890,000 short tons) from tires.

New products derived from waste tires generate more economic activity than combustion or other low multiplier production, while reducing waste stream without generating excessive pollution and emissions from recycling operations:

- Construction materials. Entire homes can be built with whole tires by ramming them full of earth and covering them with concrete, known as earthships. They are used in civil engineering applications such as sub-grade fill and embankments, back-fill for walls and bridge abutments, sub-grade insulation for roads, landfill projects, and septic system drain fields. Tires are also bound together and used as different types of barriers such as: Collision reduction, erosion control, rainwater runoff, blasting mats, wave action that protects piers and marshes, and sound barriers between roadways and residences.

- Artificial reefs are built using tires that are bonded together in groups, there is some controversy on how effective tires are as an artificial reef system, an example is The Osborne Reef Project which has become an environmental nightmare that will cost millions of dollars to rectify.

1 ton bags of crumb rubber.

- The process of stamping and cutting tires is used in some apparel products, such as sandals and as a road sub-base, by connecting together the cut sidewalls to form a flexible net.

- The markets predicted by the 2003 report were: Tire derived fuel (TDF) using 130 million tires, civil engineering projects using 56 million tires, ground rubber turned into molded rubber products using 18 million tires, ground rubber turned into rubber-modified asphalt using 12 million tires, Exported items using 9 million tires, cut, stamped and punched products using 6.5 million tires, and agricultural and miscellaneous uses 3 million tires.

- Shredded tires, known as Tire Derived Aggregate (TDA), have many civil engineering applications. TDA can be used as a back-fill for retaining walls, fill for landfill gas trench collection wells, back-fill for roadway landslide repair projects as well as a vibration damping material for railway lines.

- Ground and crumb rubber, also known as size-reduced rubber, can be used in both paving type projects and in mold-able products. These types of paving are: Rubber Modified Asphalt (RMA), Rubber Modified Concrete, and as a substitution for an aggregate. Examples of rubber-molded products are carpet padding or underlay, flooring materials, dock bumpers, patio decks, railroad crossing blocks, livestock mats, sidewalks, rubber tiles and bricks, movable speed bumps, and curbing/edging. The rubber can be molded with plastic for products like pallets and railroad ties. Athletic and recreational areas can also be paved with the shock absorbing rubber-molded material. Rubber from tires is sometimes ground into medium-sized chunks and used as rubber mulch. Rubber crumb can also be used as an infill, alone or blended with coarse sand, as in infill for grass-like synthetic turf products such as Field-turf.

- Steel mills can use tires as a carbon source, replacing coal or coke in steel man-ufacturing.

- Tires are also often recycled for use on basketball courts and new shoe products.

Closeup of shredded tires.

Tire Pyrolysis

The pyrolysis method for recycling used tires is a technique which heats whole or shred-ded tires in a reactor vessel containing an oxygen-free atmosphere. In the reactor the rubber is softened after which the rubber polymers break down into smaller molecules. These smaller molecules eventually vaporize and exit from the reactor. These vapors can be burned directly to produce power or condensed into an oily type liquid, general-ly used as a fuel. Some molecules are too small to condense. They remain as a gas which can be burned as fuel. The minerals that were part of the tire, about 40% by weight, are removed as solid ashes. When performed properly, the tire pyrolysis process is a clean operation and produces little emissions or waste; however, concerns about air pollu-tion due to incomplete combustion as is the case with tire fires has been documented.

The properties of the gas, liquid, and solid output are determined by the type of feed-stock used and the process conditions. For instance whole tires contain fibers and steel. Shredded tires have most of the steel and sometimes most of the fiber removed. Pro-cesses can be either batch or continuous. The energy required to drive the decomposi-tion of the rubber include using directly fired fuel (like a gas oven), electrical induction (like an electrically heated oven) or by microwaves (like a microwave oven). Sometimes a catalyst is used to accelerate the decomposition. The choice of feed-stock and process can affect the value of the finished products.

The historical issue of tire pyrolysis has been the solid mineral stream, which accounts for about 40% of the output. The steel can be removed from the solid stream with mag-nets for recycling. The remaining solid material, often referred to as "char", has had little or no value other than possibly as a low grade carbon fuel. Char is the destroyed remains of the original carbon black used to reinforce and provide abrasion resistance to the tire. The solid stream also includes the minerals used in rubber manufacturing. This high volume component of tire pyrolysis is a major impediment, although this theme continues to be a source of innovation.

Repurposing

Aside from recycling old tires, the old tire can be put to a new use. Old tires are sometimes converted into a swing for play. The innovative use allows for an easy way to find a purpose for an existing old tire not suitable for road use.

Used tires are also employed as exercise equipment for athletic programs such as American football. One classic conditioning drill that hones players' speed and agility is the "Tire Run" where tires are laid out side by side, with each tire on the left a few inches ahead of the tire on the right in a zigzag pattern. Athletes then run through the tire pattern by stepping in the center of each tire. The drill forces athletes to lift their feet above the ground higher than normal to avoid tripping. Other athletic uses include tire flipping (tractor or large truck tires typically used) or for upper cardio conditioning by hitting a tire repetitively with a sledge hammer.

Three boys are on a tire swing.

Car tires as seats in Thailand.

Re-purposed tires can also be harnessed as an affordable alternative building material used in the framework of rammed Earth thermal mass dwellings. This is beneficial across scales of production such as individually sustainable housing.

Rows of stacks of tyres are often used as barriers in motor racing circuits as a method of dissipating kinetic energy over a longer period of time during a crash, comparatively to striking a less malleable material such as a concrete or steel wall.

Many cattle farmers re-purpose old tractor tires as water troughs for their cattle by placing them over natural springs or by piping stream water into them. These tires contain the water and allow it to pool for the cattle without any additional interaction from the farmer. Most farmers also include a drainage pipe near the top or in the center of the tire so excess water can drain off to prevent overflow and erosion around the outside of the tire where the cattle would be.

Environmental Concerns

Due to their heavy metal and other pollutant content, tires pose a risk for the leaching of toxins into the groundwater when placed in wet soils. Research has shown that very little leaching occurs when shredded tires are used as light fill material; however, limitations have been put on use of this material; each site should be individually assessed determining if this product is appropriate for given conditions.

Tire art.

For both above and below water table applications, the preponderance of evidence shows that TDA (tire derived aggregate, or shredded tires) will not cause primary drinking water standards to be exceeded for metals. Moreover, TDA is unlikely to increase levels of metals with primary drinking water standards above naturally occurring background levels.

BOTTLE RECYCLING

Bottles are able to be recycled and this is generally a positive option. Bottles are collected via kerbside collection or returned using a bottle deposit system. Bottlerecycle.org/ reveals that just 14% of all plastic packaging is recycled globally PET bottles production is predicted to grow by about 5% a year. Currently just over half of plastic bottles are recycled globally About 1 million plastic bottles are bought around the world every minute and only about 50% are recycled.

Glass Bottles

Glass bottles are fully recyclable, either to be reused as bottle or to be melted and reused as glass, resulting in a saving of energy and raw materials. Recycling one glass bottle can save enough energy to power a computer for 25 minutes.

PET Bottles

PET bottles are mostly recycled as a raw material. In many countries, PET plastics are coded with the resin identification code number "1" inside the universal recycling symbol, usually located on the bottom of the container.

HDPE Bottles

HDPE is commonly used in bottles, particularly bottles (or jugs) of milk. Recycling code 2 is applicable. In the US, only about 30-35% of HDPE bottles are recycled.

Legislation

Container deposit legislation are laws passed by city, state, provincial, or national governments. They require a deposit on bottles to be collected when sold and reimbursed when returned. In May 2018 the Israeli Ministry Impose EUR 12 m Fine on Bottle Manufacturers and Importers that Didn't Meet Collection Targets.

Environmental Comparisons

Many potential factors are involved in environmental comparisons of returnable vs non-returnable systems. Researchers have often used life cycle analysis methodologies to balance the many diverse considerations. Often the comparisons show benefits and problems with all alternatives. It helps provide an objective view of a complex subject.

Reuse of bottles requires a reverse logistics system, cleaning and, sanitizing bottles, and an effective Quality Management System. A key factor with glass milk bottles is the number of cycles of uses to be expected. Breakage, contamination, or other loss reduces the benefits of returnables. A key factor with one-way recyclables is the recycling rate: In the US, only about 30-35% of HDPE bottles are recycled.

DRUG RECYCLING

Drug recycling is the idea that health care organizations or consumers with unused drugs can transfer them in a safe and appropriate way to another consumer who needs them. This would happen through a specialized pharmacy or medical organization which would oversee and mediate the process.

It is traditional to expect that consumers get prescription drugs from a pharmacy and that the pharmacy got their drugs from a trusted source. In a drug recycling program, consumers would access drugs through a less regulated supply chain and consequently the quality of the drugs could be lower. Advocates and opponents of drug recycling programs differ in judging the risk of the less regulated supply chain lowering the quality of drugs which reach the consumer.

Various regional governments in the United States offer drug recycling programs. As of 2010, Canada had fewer drug recycling programs than the United States.

Various organizations and governments are experimenting with different ways to manage drug recycling programs. Many drug recycling programs only take drugs from professionals, and never from patients. Drug recycling programs are not designed to be useful for consumers seeking to dispose of drugs. These programs also are not intended to lessen the environmental impact of pharmaceuticals and personal care products. Typically they only accept medication which has not expired and which is in unopened packaging. When re-using recycled drugs, they only go to specific pharmacies which are prepared to address the special requirements of participating in a recycling program. Usually, drug returns happen without financial compensation. SIRUM, an acronym for "Supporting Initiatives to Redistribute Unused Medicine", is a nonprofit organization which advocates for drug recycling.

Need for Medicine Recycling

The high cost of medicines (both to consumers and third parties) is a problem that seems to be getting worse rather than better. The new Medicare prescription benefit program does not address the problem of acquisition price, since it prohibits the federal government from directly negotiating with the pharmaceutical manufacturers. "Every time Medicare expands a benefit, patients come out of the woodwork." Thus, with little or no pressure on the purchase price and more affordability (at least for seniors), the overall national expenditure devoted to medication will continue to rise. Furthermore most of the population is not covered by Medicare.

Many patients, do not take their medications as prescribed because of cost; as a result, more than half reported subsequent health problems. Severe disability, poor health, low income, lack of insurance, and a high number of prescriptions increase the odds of cost-related noncompliance. No doubt, the above is but the tip of an iceberg; many Americans are not able to afford medication vital to their health and well-being. They either completely do without, skip doses, or fail to refill medication prescribed for chronic medical conditions — all because of the perceived high cost of medicines.

Rationale for Medicine Recycling

Although other countries have opted for pharmaceutical price controls and curtailing patent protection, such actions run counter to American free-market ideology and political realities. Pharmaceutical manufacturers point out that their industry spent more than $30 billion on drug research and development in 2001, as compared with the total National Institutes of Health operating budget of $20 billion. There is also evidence of a competitive marketplace; in the first year that a medicine goes off patent, its generic equivalent sells for about half the price of the brand-name drug. These and other arguments are sure to be heard by politicians, since pharmaceutical industry political contributions have escalated dramatically and doubled between 1998 and 2002. During the 2002 election campaign, the drug industry contributed $21 million to federal political campaigns for parties and candidates. According to the Center for Responsive Politics, that makes the group the largest contributor.

Drug benefit managers and preferred drug lists or formulary restrictions provide some pressure to keep prices down, but that strategy fails for unique, brand name compounds without any available equivalent. The marketplace seems to be missing a way of enhancing competition for medicines still under patent. One possibility is to allow competition from exactly the same molecule made by the same manufacturer while still under patent protection. Reimporting medicine made in the United States from Canada or Mexico is an example of such a gambit. Unfortunately, the drug makers are already undercutting the effectiveness of this tool. For example, GlaxoSmithKline (GSK) announced in January 2003 that it would no longer sell products to Canadian pharmacies and wholesalers that market GSK drugs over the Internet to other countries. Pfizer, the world's biggest drug maker, following along after GSK (as well as AstraZeneca and Wyeth), is carefully monitoring the orders of Canadian pharmacies it believes are exporting to the United States and now limiting orders to those pharmacies. The manufacturers have the capacity to know who is exporting, since data-tracking companies and drug benefit managers monitor individual doctors' prescriptions. When pharmacy orders spike at a particular pharmacy in the absence of a similar jump in nearby doctors' prescriptions, manufacturers are alerted.

Another possible supply for expensive medication still under patent exists. It is the unused supply that patients are left with when they stop or change medicines. Could we somehow get those unused tablets and capsules to those in need at a reasonable price? Why not recycle widely used expensive medications like atypical antipsychotics, where each tablet may cost $7–$14 each, or Zofran (an antinausea medication often used for cancer patients undergoing chemotherapy), which costs close to $30 per tablet? How about an effort to recycle Gleevec capsules, which cost approximately $27,000 per year for the treatment of a single patient with chronic myelogenous leukemia, a price based in part on the price of interferon, the next best available drug treatment for that disease.

Recycling may also offer a free-market solution to the problem of artificially stimulated demand for a product often paid for by others. The current emphasis on marketing is illustrated by a Families USA analysis of the pharmaceutical industry's brand-name leading companies' 2001 annual financial statements. Brand-name drug makers in the United States employ 81% more people in marketing than in research. Whereas marketing staffs increased by 59% between 1995 and 2000, research staffs declined by 2%. On average, the 9 leading companies spent 11% of revenue on research and development compared with 27% of revenue on marketing, advertising, and administration in 2001.

However, with a recycling program in place, overmarketing expensive, brand-name medicines might backfire on the manufacturer. High initial sales based on unrealistic patient expectations would result in a large supply of abandoned medication available to re-enter the market place at a low price, thereby undercutting sales of newly manufactured tablets. For example, the current blitz of advertising for a variety of brand medications to treat male erectile dysfunction will likely end up with many unused tablets in patients' bathroom cabinets as the medicines do not always work for everyone, nor do they automatically result in happy marriages. Currently, the huge collection of oversold, unused tablets does not adversely affect future sales to other users. However, with medication recycling in place, those unused tablets for erectile dysfunction will come back into the marketplace, offering a cheap alternative to newly produced tablets.

Supply Sources for a Medicine Recycling Program

Nursing homes and other healthcare facilities are not the only possible sources of recyclable medicines. A more important source may be medication in consumers' homes that is unused. Many prescriptions are frequently switched or simply stopped in midstream by prescribers. Leslie and Rosenheck documented the switching phenomenon in a recent article tracking the prescription of antipsychotic medications within the Department of Veterans Affairs. Of the 21,873 patients with schizophrenia who were on stable 3-month prescriptions of any antipsychotic medication, 25% had their medications switched during the next year.

Patients are also likely to stop medication on their own. Not more than 50% of patients adhere to a chronically prescribed antipsychotic medication, typical or atypical in type, for one year. Antidepressants, another costly and widely used medication class, are discontinued even more frequently. As we move away from the mental health arena, adherence is still a major problem. The cholesterol-lowering statins, which, like atypical antipsychotic medications and antidepressants, are prescribed most often for chronic use, show a similar pattern of premature discontinuation, leaving much unused, expensive medicine in consumer hands. Adherence rates for all medications prescribed for asthma range from 30% to 70%.Mail order prescriptions for 90 days (rather than the 30-day supplies usually available from local pharmacies) have further increased the amount of unused medication in consumers' hands.

Expiration Dates

"Expired medication" also represents a potentially large source of supply of medicine for a recycling program. As things stand now, expiration dates get a lot of emphasis. For instance, there is a campaign, cosponsored by some drug retailers, that urges people to discard tablets or capsules when they reach the date on the label. It turns out that the date on the label, however, is often much earlier than the official expiration date. Pharmacists are required to put a "beyond-use" date on prescriptions, which is either the manufacturer's expiration date or 1 year from the date the drug is dispensed, whichever is earlier. The rationale is that containers into which dosage forms are repackaged may not have the integrity of the original package.

However, not only the beyond-use date, but the official drug expiration date itself is usually determined conservatively, and very expensive medication is being wasted. Data from the Department of Defense/US Food and Drug Administration (FDA) Shelf

Life Extension Program, which tests the stability of drug products past their expiration date, showed 84% of 1122 lots of 96 different drug products stored in military facilities in their unopened original container would be expected to remain stable for an average of 57 months after their original expiration date. Some US Army studies on Valium, for example, show that the drug is very stable and completely safe and effective for up to 8 years after manufacture. Tablets of ciprofloxacin, an expensive antibiotic, were found completely safe and effective when tested 9.5 years after the expiration date. Theophylline, in tablet form, shows 90% stability even after 30 years beyond the expiration date. Such stability is not reflected in the manufacturer or pharmacy dating about when tablets or capsules must be discarded. In general, although published data are not available for all medicines, consultants believe that most drugs stored under reasonable conditions retain at least 70% to 80% of their potency for at least 1 to 2 years after the expiration date, even after the container has been opened (nb: current US Pharmacopoeia [USP] standard is generally 90% potency). With new individual pill packaging techniques, it is highly likely that USP acceptable potency would be the norm over that same time period.

The only report of human toxicity that may have been caused by chemical or physical degradation of a pharmaceutical product is a disputed article alleging renal tubular damage associated with use of degraded tetracycline. The lack of other reports of toxicity from expired medication is reassuring, but the topic of out-of-date medication toxicity is not a well-researched issue.

New Technology for Assuring Safety and Potency of Recycled Medicine

If we have already learned how to recycle everything from newspapers to soda cans and rubber tires, eyeglasses and human organs such as corneas and kidneys, then why not medicines? Medicines would not even have to be remanufactured, just verified. Many tablets or capsules already are packaged individually in tamper-proof wrappings with the name and date of expiration stamped on each one. Such blister packaging is already routine for many over-the-counter medicines and physician samples.

Bar coding of identifying information on the individual pill or capsule wrapper would also be helpful, and we may be closing in on that possibility. The FDA just passed a final rule requiring bar coding on all medication labels (ie, immediate container label and the outside container or wrapper) for all products that might be dispensed in a hospital or other institutional setting. The labels would have to be imprinted with a bar code containing the National Drug Code number that identifies each drug and its dosage form and strength. The requirement would not preclude the inclusion of other information, such as the drug's lot number and expiration date. Bar codes on the medication label would allow for computer scanning at the bedside in hospitals and nursing homes to verify that drug and dosage matched the patient bar code imprinted on patient-specific bracelets — thereby guarding against medication errors.

Unfortunately, since the FDA did not mandate unit dose packaging along with the bar coding, there is still a possibility of bedside errors. Even more ominously, the additional cost of requiring bar codes may cause a decline in the use of unit dose packaging. So, what looks like a very good safety measure may not be such a clear winner. It is clear that bar coding of each unit dose of medicine would be what would be required.

The FDA is also assessing the extent to which new technologies in development, such as color-shifting inks and images (seen only when the package is turned in a certain way) could defeat unlawful copying and counterfeiting. Micro-bar-code tags, special perfumes, embedded threads, and implanted radio frequency chips within an individual tablet or capsule blister package might be utilized to establish authenticity and integrity of medicines. While most of these new approaches have not yet been fully developed, implemented, and tested, they hold the promise of a more secure drug distribution system in the years ahead. Not only do these technologies have the potential to make recycling more feasible than previously thought, they will come about anyhow to counter the growing problem of counterfeit drugs.

Additional new developments in packaging address the many compounds that are moisture sensitive, for which exposure to sustained high temperatures and humidity results in significant degradation. However, when a very moisture sensitive compound (eg, PGE-7762928) is packaged in cold-form aluminum blisters, even harsh conditions — 6 months at 40° Centigrade and 75% humidity — results in 100% preservation of physical and chemical stability. Not all blister packaging works as well. In the same study, with a baseline of 82% sustained activity of the compound when unpackaged, there was 84% stability when stored in polyvinyl chloride blisters, 91% in cyclic olefin blisters, 97% in aclar blisters, and 99% in a high-density polyethylene bottle with a foil induction seal. Other studies show that mathematical models can predict moisture uptake by packaged pharmaceutical products during storage under different conditions.

Oxidation is another common way for medicines to lose potency, and there are great complexities in how that may happen, depending up on the compound involved. Still, modern individual dose packaging, either by the manufacturer or wholesaler, can help enormously. Blister packaging, especially in opaque foil blisters, works extremely well. It is even feasible to blister package under nitrogen or argon to reduce the oxygen in the package to begin with. Alternatively, blister packaging with oxygen absorption capacity (eg, self-activated, iron-based oxygen scavengers) could be utilized such that single doses can be maintained under anaerobic conditions. Plastics containing oxygen scavengers have also become available and result in what some have termed "active packaging." Another new tool, high-pressure processing, already useful in protecting food and extending its shelf life, might be used for the stabilization of heat-labile, fragile pharmaceuticals.

As medicines have risen in price, the differential cost for more expensive protective blister packaging becomes a relatively trivial expense. When one is talking about tens of dollars or more for each pill or capsule, a few cents investment is clearly worthwhile to maintain the integrity of a product long enough (even under adverse storage conditions) to be able to reach a needy patient. Either the FDA or Congress might mandate the use of individual tablet/capsule packaging and identification, or wholesalers or large pharmacy chains might do it for all new prescriptions, expensive items, and any items deemed eligible for return credit and recycling.

Environmental Impact

Although a clean environment is widely championed, we hardly ever talk about pharmaceutical pollution. Recently, pharmaceuticals have been detected in surface water, ground water, and drinking water. Furthermore, resistant bacteria may be selected in the aeration tanks of sewage treatment plants by the antibiotic substances present. Efforts to make sure terrorists do not contaminate our drinking water are in force, herbicide and pesticide levels are examined, but pharmaceuticals abound, and while they are likely safe for humans directly, this may not be so when they are unleashed to work indiscriminately and over time in our environment. One advantage of a pharmaceutical recycling program would be the safe disposal of whatever is unneeded or truly out of date. A 1996 report of how expired medications are being disposed of found that 1.4% of respondents returned medications to a pharmacy, 54% disposed of medications in the garbage, 35.4% flushed medications down the toilet or sink, 7.2% did not dispose of medications, and 2% related they used all medications before expiration.

Another source of environmental pollution, pharmaceutical contamination by way of intact molecules (or active metabolites) passing through urine and feces, may actually be a larger problem, but it is one far removed from the issue of medication recycling.

Public Health Benefits of Recycling Medicines

Unfortunately, when consumers do not properly dispose of medicines, there is also the problem of accidental or, more likely, purposeful overdosage. Suicide is a major public health problem, and the most likely form suicide attempts take is with overdosages of medicine. Whereas many physicians treating suicide-prone patients limit prescription quantities of potentially lethal medicines, that does nothing for the stored bottles of medicines that reside in consumers' homes. Lethal quantities of residual pills in medicine cabinets invite impulsive suicidal action, as well as accidental ingestion by children.

Another benefit from a recycling program would be information about why patients prematurely stop their medication. Information compiled about which medicines seemed to get turned in most frequently is sorely needed to learn about real-world medication effectiveness, side-effects, and adherence. Such data would highlight problems with specific medicines quickly and would certainly be of interest both to pharmaceutical manufacturers and the FDA. Recycling pharmacists might also notify prescribers about the early return of vital medication and directly educate patients about the need for continuing medication for chronic indications such as diabetes, hypertension, dyslipidemia, major depression, schizophrenia, AIDS, etc.

RECYCLING PROCESSES FOR PHOTOVOLTAIC MODULES

Photovoltaic (PV) solar modules are designed to produce renewable and clean energy for approximately 25 years. The first substantial PV installations happened in the early 1990s and since early 2000s solar PV electricity distribution has grown extremely fast.

The cumulative worldwide PV generation capacity reached 302 GW in the end of 2016 and the predominant technology (90% of the market) is crystalline silicon (c-Si) cells. Also, during the last years there were several advances on renewable energy in general, including significant price decline and a constant increase in attention to environmental impacts from energy sources. Furthermore, the International Technology Roadmap for Photovoltaic (ITRPV) prediction for the installed PV capacity in 2050 is 4500 gigawatts.

As a result of the increase in the global market for PV energy, the volume of modules that reach the end of their life will grow at the same rate in the near future. At the end of 2016, the cumulative global PV waste reached 250,000 metric tons, while it is expected that by 2050 that figure will increase to 5.5–6 million tons.

Much PV waste currently ends up in landfill. Given heavy metals present in PV modules, e.g. lead and tin, this can result in significant environmental pollution issues. Furthermore, valuable metals like silver and copper are also present, which represents a value opportunity if they can be recovered. Hence, the landfill option cerates additional costs and it does not recover the intrinsic values of the materials present in the PV modules.

Hence, methods for recycling solar modules are being developed worldwide to reduce the environmental impact of end-of-life modules and to recover some of the value from old PV modules. However, current recycling methods are mostly based on downcycling processes, recovering only a portion of the materials and value, so there is plenty of room for progress in this area. Moreover, currently only Europe has a strong regulatory framework in place to support recycling, but other countries are starting to build specific frameworks related to PV waste. It's clear that sustainable development of the PV industry should be supported by regulatory frameworks and institutions across the globe, which is not the case at the moment. There must be adequate management policies for photovoltaic modules when they reach their end-of-life (EoL) or when they are not able to produce electricity any longer.

The European Union (EU) provides a legislative framework for extended producer responsibility of PV modules in European scale through the Waste Electrical and

Electronic Equipment (WEEE) Directive 2012/19/EU. The main objectives of this policy are to preserve, protect and improve the quality of the environment, to protect human health and to utilize natural resources prudently and rationally. Since February 2014, the collection, transport and recycling of PV modules that reached their EoL is regulated in every EU country.

On the other hand, countries with fast expanding PV markets such as China, Japan, India, Australia and USA still lack specific regulations for EoL PV modules. These countries treat PV waste under a general regulatory framework for hazardous and non-hazardous solid waste or WEEE. However, there are some exceptions.

In 2012 the Japanese government introduced a "feed-in tariff" that guaranteed the rate for electricity generated from renewable energy and exported to the grid, which supported rapid growth of solar module installation in the country. Once all the installed capacity starts reaching EoL (within 20–30 years) they will create a significant waste problem for Japan. Manufacturers, importers and distributors of photovoltaic modules have been invited to provide information on the chemical substances contained in the product and to inform the waste disposal companies.

In USA, some states go beyond the Resource Conservation and Recovery Act which regulates hazardous and non-hazardous waste management. California, for example, has additional threshold limits for hazardous materials classification based on the Senate Bill 489 that categorizes end-of-life PV modules as Universal Waste (facilitating easy transport). This bill is currently pending United States Environmental Protection Agency approval.

In Australia, governments have recognized the significance of guaranteeing that regulations are in place to deal with the PV waste issue. Ministers agreed that the state of Victoria would lead innovative programs that seek to reduce the environmental impacts caused throughout the lifecycle of photovoltaic systems. These efforts are part of an industry-led voluntary product management arrangement to address the potential emerging risks of PV systems and their waste. PV modules are listed under the National Product Administration Act to signal the intention to consider a scheme to deal with such waste.

The non-inclusion of PV residues in waste legislation in some countries is due to different reasons. Solar modules have a lifespan of up to 25–30 years and so there has been limited interest in investigating the aspects of EoL so far. Moreover, the quantity of this type of waste is still considered insignificant compared to the quantity of other WEEE, which currently makes setting up specific recycling plants for solar modules uneconomical. In addition, the definition of mandatory requirements for EoL treatment could still be an obstacle to the effective acceptance of these recycling processes. Because of that, there should be a continuous focus on scientific evidences on the potential impacts and benefits related to the treatment of photovoltaic residues.

Furthermore, recycling processes for all the different PV technologies are not yet well developed. The processes are well developed for mono or multicrystalline silicon. First-Solar has an established recycling process for CdTe, but for other thin films there are still room for improvements. and are being tested and for generation 3 (new materials) the recycling technologies are not well developed yet.

Only about 10% of PV modules are recycled worldwide. The main reason for that is the lack of regulation. Actually, it has been shown that, for the current recycling technologies, silicon-based modules do not have enough valuable materials to be recovered and the cost of the recycling process is always higher than the landfill option (not considering the externalities), making recycling an economically unfavorable option. However, the prediction for 2050 is that the recoverable value could cumulatively exceed 15 billion US dollars (equivalent to 2 billion modules, or 630 GW). In addition, the recycling of solar PV modules can ensure the sustainability of the long-term supply chain, thereby increasing the recovery of energy and embedded materials and, also, reducing CO_2 emissions and energy payback time (EPBT) related to this industry.

For years, the PV industry and researchers have worked intensively in search of different types of efficient and cost-effective materials to manufacture solar PV modules and specific ways of keeping them adequately bonded to withstand several years of outdoor exposure. The modules are made to minimize the amount of moisture that can come in contact with the solar cells and their contacts while keeping manufacturing costs down. The current standard c-Si module is bonded using two layers of EVA to bond the layers together. Because of that, recycling solar modules is a relatively complex task, since these materials need to be separated. Once the materials/layers of a solar module can be separated, metals such as lead, copper, gallium, cadmium, aluminum and silicon can be recovered and reused in new products.

Originally created by PV CYCLE in 2007 and commercially available in Europe, the process of recycling mono or multicrystalline silicon modules begins with the separation of the aluminum frame and the junction boxes and then a mechanical process is used for the extraction of the remaining materials of the module (a process similar to recycling of glass or electronic waste). The problems with this process are that the value of the material recovered is low (as it is a downcycling process) and that the maximum amount of recovered materials is about 80%, which is not sufficient for future requirements, and the value of recovered materials is smaller than the original. Thin film processes are under development or near implementation in Italy, Japan and South Korea but costs are not yet competitive. Even up to 90% recovery of materials is not sufficient when compared to production costs. Lastly for recycling processes aiming to generate new materials, the aim is to keep the materials intact for reuse or direct recycling, recovering the frame, glass, tabbing and solar cells without breakages and in good condition. The recovery rates can achieve up to 95% and the materials recovered have higher commercial value. However, these processes are complex and are currently just at laboratory scale, being studied by a few research groups.

Even with the difficulty of recovering rare, toxic and valuable materials from solar modules, the recycling process has a remarkable environmental advantage. Nevertheless, the need to recycle this type of waste is imminent. The better knowledge of these technologies and growth on the waste amounts that could generate profitable outcomes has supported the development of the first PV recycling plants. Hence, PV manufacturing companies (e.g. First Solar, Pilkington, Sharp Solar, and Siemens Solar) are investing in the research on solar modules at EoL.

The challenges to design the ideal PV recycling process are many. The focus should be on the avoidance of damage to the PV cells and module materials, economic feasibility, and high recovery rate of materials that have some monetary value or are scare or are hazardous, that can be reused in the supply chain. Finally, the next step for the industry and researchers is to create module designs that are "recycling-friendly".

Photovoltaic Technologies

Crystalline Silicon Technology

Crystalline Si (c-Si) technologies dominate the current market share of PV modules (more than 90%). The aluminum back surface field (Al-BSF) is the current industry standard technology but the passivated emitter and rear cell (PERC) is gaining importance in the world market and is expected to replace the Al-BSF technology in the future. The heterojunction (HIT) cells are also expected to gain some space with predictions of 15% of the total market share by 2027. Besides that, Si-based tandem solar technologies are expected to appear in mass production after 2019.

There are different cell structures for crystalline silicon-based PV cells. The cells are electrically interconnected (with tabbing), creating a string of cells in series (60 or 72 cells are standard numbers used) and assembled into modules to generate electricity.

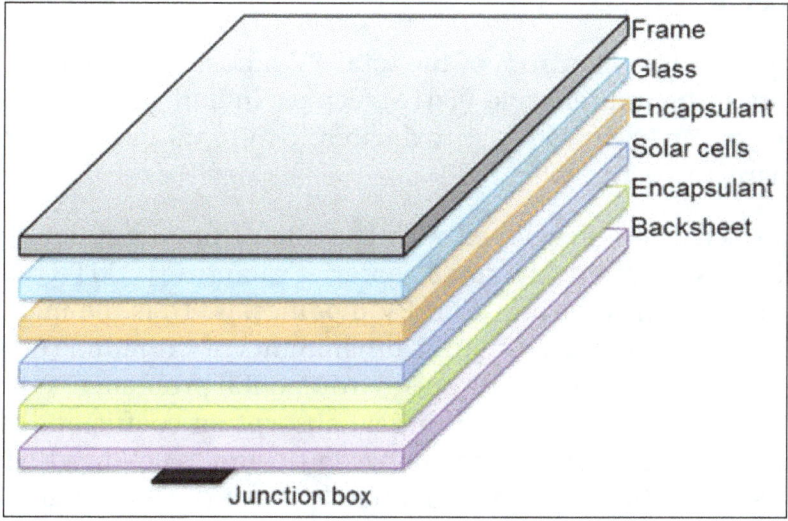

Silicon solar module basic structure.

A typical crystalline silicon (c-Si) PV module contains approximately 75% of the total weight is from the module surface (glass), 10% polymer (encapsulant and backsheet foil), 8% aluminum (mostly the frame), 5% silicon (solar cells), 1% copper (intercon-nectors) and less than 0.1% silver (contact lines) and other metals (mostly tin and lead). The rest of the components have a small percentages of the module weight.

The EU directive established recycling targets in terms of module weight and also ex-presses the intention to increase the collection rates to allow the progressive recycling of more material and less to be landfilled. Even with targets aiming for 65% recycling product weight, some of the current studied recycling processes can recycle over 80% of the weight of a PV module. However there is still incentive to improve, considering that most of the weight is from glass and frame, which are relatively easy to remove, depending on the recycling process.

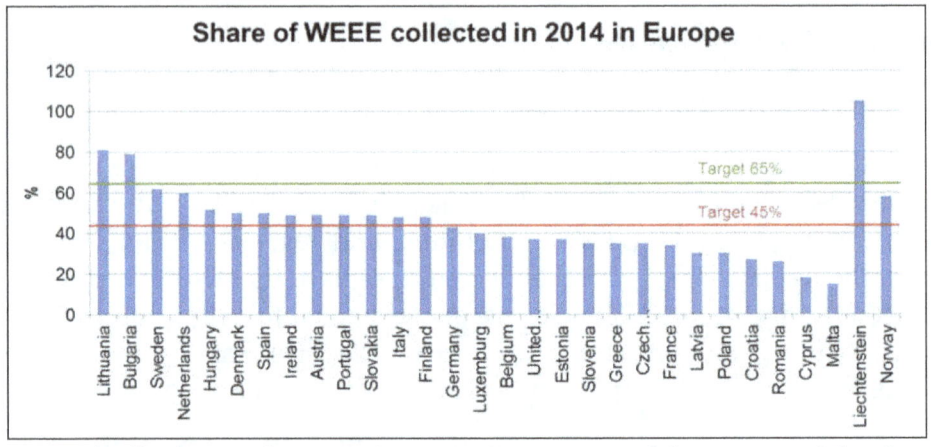

Total collection rate for WEEE in 2014 as a percentage of the average weight of EEE put on the market in the three preceding years.

Thin-film Technologies

Thin-films represent less than 10% of the total PV industry. The currently dominant technologies are cadmium telluride (CdTe), copper indium gallium selenide (CIGS) and amorphous silicon (a-Si) with, approximately, 65%, 25% and 10% of the total thin-film market share, respectively.

Thin-film solar cells were developed with the aim of providing low cost and flex-ible geometries, using relatively small material quantities. CdTe, CIGS and a-Si are the main technologies for thin-film PV modules. CdTe is the most widely used thin-film technology. It contains significant amounts of cadmium (Cd), an element with relative toxicity, which presents an environmental problem that has been stud-ied worldwide. CIGS has a very high optical absorption coefficient because it is a direct band gap material (can be tuned between 1.0 and 2.4 eV by varying the In/Ga and Se/S ratios) and efficiency of approximately 15.7 ± 0.5% for high bandgap. A-Si has low toxicity and cost but also low durability and it is less efficient compared with the other thin-film technologies. Current projections expect the a-Si module market to disappear in the near future, since they cannot compete on costs or efficiency.

Basically, thin-film modules consist of thin layers of semiconducting material (CdTe, CIGS or a-Si) deposited on a substrate (glass, polymer or metal).

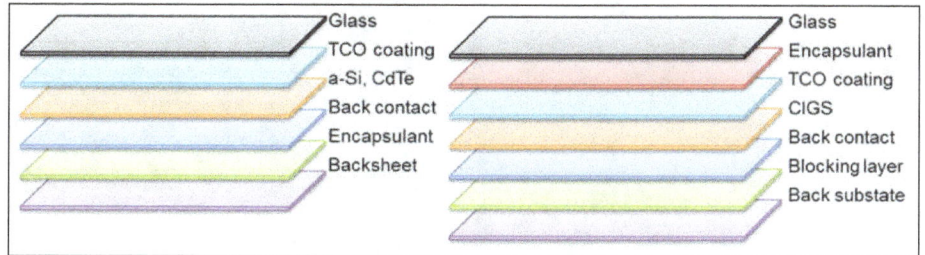

Thin-film solar module basic structures.

Photovoltaic Recycling Technologies

PV modules are largely recyclable. Materials such as glass, aluminum and semiconductors can, theoretically, be recovered and reused. Hence it is vital that consumers, industry and PV producers take responsibility for the EoL of these modules. So far, the most common methods for recycling c-Si PV modules are based on mechanical, thermal and chemical processes.

Although thin-film solar cells use far less material than c-Si cells, there are concerns about the availability and toxicity of materials such as tellurium (Te), indium (In), and cadmium (Cd), for example. Furthermore, the production processes also generates greenhouse gases emissions during some reactor-cleaning operations. Because of these issues, it is very important to focus on the recycling of PV modules for all the technologies.

PV Cycle is a not-for-profit organization which goal is to manage PV waste through their waste management programme for solar PV technologies. PV Cycle was the first to establish a PV recycling process and PV waste logistics throughout the EU. In 2016 their process of recycling PV achieved a record recycling rate of 96% for c-Si PV modules (fraction of solid recycled), which is a percentage that surpasses the current European WEEE standards. The process begins with the removal of the cables, junction box and frame from the PV module. Then, the module is shredded, sorted and separated. The separation of the materials allows them to be sent to specific recycling processes associated with each material. The summary of this process is shown in figure.

Summary of PV cycle recycling process for c-Si modules.

FirstSolar developed a recycling process for CdTe modules. The company manages the collection and transportation of EoL modules to the recycling centre; however, the recycling process itself must be financed. This is made by setting aside funds by the company itself at the time of the module sale, which also happens with WEEE. The summary of this process is shown in figure.

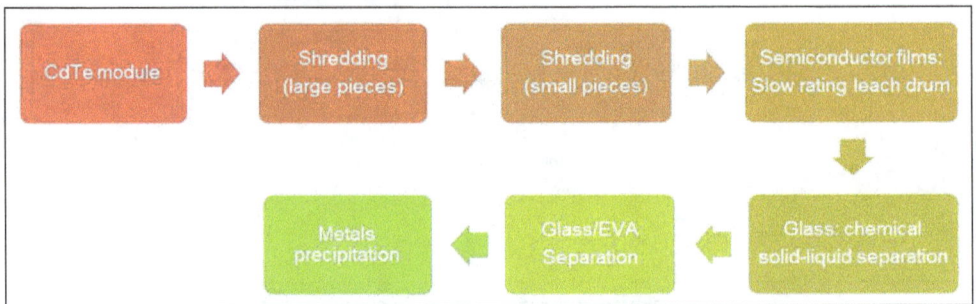

Summary of first solar recycling process for CdTe modules.

The recycling process starts with the shredding of the modules into large pieces and subsequently in to small fragments (5 mm or less) by a hammer mill. During the next 4–6 h the semiconductor films are removed in a slow leaching drum. The remaining glass is exposed to a mixture of sulfuric acid and hydrogen peroxide aiming, to reach an optimal solid–liquid ratio. After that process, the glass is separated again. The next step is to separate the glass from the larger ethylene vinyl acetate (EVA) pieces, via a vibrating screen. The glass is cleaned and sent to recycling. Sodium hydroxide is used to precipitate the metal compounds, after which they are sent to another company where they can be processed to semiconductor grade raw materials for use in new solar modules. This process recovers 90% of the glass for use in new products and 95% of the semiconductor materials for use in new solar modules.

Also, for recycling CdTe modules, ANTEC Solar GmbH designed a pilot plant with a similar technology to the First Solar process. It starts with a physical fragmentation of the modules. After that, these small pieces are exposed to an atmosphere containing oxygen at 300°C. These conditions result in the delamination of the EVA. Subsequently, these fragments are taken to a 400°C atmosphere containing chlorine gas which causes an etching process. This step of the process generates $CdCl_2$ and $TeCl_4$ that are condensed and precipitated afterwards. The summary of this process is shown in figure.

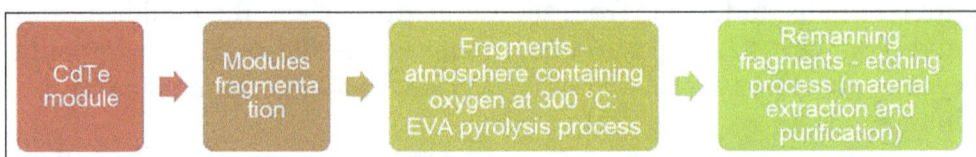

Summary of ANTEC solar GmbH recycling process for CdTe module.

A company that has a well stablished c-Si recycling process is the SolarWorld. This company started recycling in 2003 with a pilot plant using a thermal process. Today, the take-back of modules is organized via a "bring-in" system. Their process is based on a thermal process, which starts by pyrolising the modules. During this process, the plastic components are burnt at 600°C. The solar cells, glass and metals are separated manually after that. The glass and some metals are sent to other companies for recycling and the solar cells can be turned into wafers again. The outcomes of this process are the recovery of more than 84% of the module weight, being 90% of the glass and 95% of the semiconductor materials. This process can recover up to 98% unbroken cells depending on the conditions of the module and the thickness of the cells. The summary of this process is shown in figure.

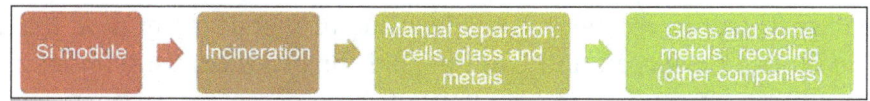

Summary of SolarWorld recycling process for Si modules.

A pilot project was funded by the Japanese Government via the New Energy and Industrial Technology Development Organization (NEDO). The recycling process for Si or CIS is based on pyrolysis of the polymers in a furnace. The process starts with the removal of the frames and the backsheet foil before the thermal process begins. After that, for CIS only, the EVA resin is burned and the CIS layer is grated. For the c-Si modules, the semiconductor materials are recovered as well as the glass cullet. The summary of these processes is shown in figure.

Summary of NEDO recycling process for Si modules (pilot project).

In 2014 the Environment Ministry of Japan, through NEDO, together with private companies, began working on new technologies to pry the PV modules apart. The new technology appeared to solve a clear problem, the firm attachment of the glass and the cells to the EVA, and the consequent difficulty to separate them simply by smashing them to pieces and sorting them out.

NPC incorporated is one of the companies that make solar module recycling equipment. The process, called the "hot knife method", can separate the cells of a module from the glass in about 40 seconds. It places the module between two rollers, which move it along and hold it steady until it runs into a 1 meter-long steel blade ("hot knife") that is heated to 180–200°C and slices the cell and the glass apart.

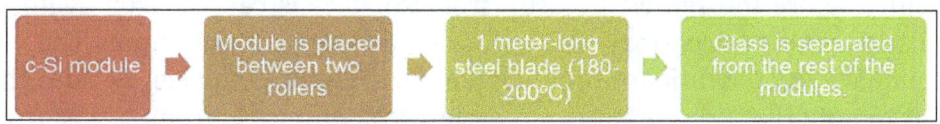

Summary of "hot knife" recycling process for PV modules.

The scrap glass can be sold for 0.5–1 Yen/kg. At that price, the 10–15 kg of glass in a solar module is worth about 15 Yen (approximately 0.14 US D). Their goal was to develop a recycling technology that can cost less than 5 yen/watt (1000 yen for a 200-watt module, not including transportation cost) by the end of April 2018, which they already did by January 2018. Furthermore, some innovative treatment processes for recycling PV solar modules have been developed.

Loser Chemie has some collection points from where they gather several types of photovoltaic systems (c-Si, CdTe, CIGS and GaAs). The company has developed and patented original processes using mechanical and chemical treatment to recycle solar cells. The first step is to crush and separate the materials mechanically. In the next stage, they use chemical treatment to recover the semiconductor metals. After that, the aluminum metallisation is also recovered and can be used for producing wastewater treatment chemicals as aluminum oxide. The summary of these processes is shown in figure.

Summary of loser Chemie recycling process for PV modules (pilot project).

Reclaim PV has teamed up with major solar module manufacturers who distribute in Australia and is refining its processes. The company is developing a process of reclaiming efficient cells from damaged solar modules. Their cell recycling system is able to extract efficient components (but not unbroken cells) from end-of-life solar modules in order to develop new green products or be reintroduced into the PV industry as new solar modules.

Environmental Aspects

Several studies have analyzed the impacts of recycling processes for PV modules on the environment. There are advantages and disadvantages of the different methods, considering all the stages, from the collection of the PV modules to the end of the recycling process.

An environmental study made for the European Full Recovery End-of-Life Photovoltaic (FRELP) project showed that environmental impacts from c-Si recycling processes come from plastic incineration and some chemical and mechanical treatments (sieving, acid leaching, electrolysis, and neutralization) for the recovery of metals.

Additionally, before the recycled silicon from solar cells can be used again, further chemical treatment is necessary, as well as for silver and aluminum. The chemical treatments have the potential of producing environmental impacts. Besides that, it is important to note that no process can recycle 100% of recovered materials from solar modules yet.

Nevertheless, for the PV Cycle c-Si recycling process it was shown that there is a significant decrease in Global Warming Potential impacts (up to 20% compared to the process of making cells) and for CdTe modules, there is and environmental benefit from the glass and copper recycling.

When comparing c-Si recycling and landfill EoL scenarios it was found that the environmental impacts from the recycling process are lower than for landfill, assuming that the recycled resources go back to the PV cells and modules manufacturing. These results considered that the recycling process involving dismantling, remelting, thermal and chemical treatments.

It can be seen that there are opportunities and challenges related to PV recycling processes. Although it was already show that there are environmental benefits, the recycling methods still need to improve in order to achieve better recovery rates and work on the transportation issues.

Economic Aspects

The recovery of valuable materials during the recycling of PV modules can have great economical value. The extraction of secondary raw material from EoL PV modules, if made in an efficient way, can make them available to the market again.

Attention has been paid particularly to silver. PV modules that reach their EoL will build up a large stock of embodied raw materials, which can be recovered and become available for other uses or even for solar cells again. However, this will not occur before 2025, according to some forecasts.

The ITRPV predicts that, by 2030, the total material value recovered from PV recycling can reach USD 450 million. With this amount it is possible to produce 60 million PV modules (18 GW), which would be approximately 33% of the 2015 production. Considering Si, up to 30,000 t of silicon can theoretically be recovered in 2030, which is the amount of silicon needed to produce approximately 45 million new modules. Considering a polysilicon current prices at USD 20/kg and a recovery rate from commercial recycling processes of 70% this is equivalent to USD 380 million.

References

- Kane, Raymond; Sell, Heinz [editors] (2001). Revolution in lamps: a chronicle of 50 years of progress (2nd ed.). Lilburn, GA: Fairmont Press. ISBN 0-88173-378-4

- Uses-recycled-oil, recycling-your-oil, used-oil-recycling, protection: environment.gov.au, Retrieved 26 April, 2020

- Buekens, A.; Yang, J. (2014). "Recycling of WEEE plastics: A review". The Journal of Material Cycles and Waste Management. 16 (3): 415–434. doi:10.1007/s10163-014-0241-2

- Recycling-used-oil-filters-shop, recycling-resources: steelsustainability.org, Retrieved 20 August, 2020

- Bernardes, A. M.; Espinosa, D. C. R.; Tenorio, J. A. S. (3 May 2004). "Recycling of batteries: a review of current processes and technologies". Journal of Power Sources. 130 (1–2): 291–298. Bibcode:2004JPS...130..291B. doi:10.1016/j.jpowsour.2003.12.026. ISSN 0378-7753

4

E-waste Recycling

E-waste refers to the discarded electronic devices. It includes electronics wastes such as computer monitors, printers, scanners, keyboards, circuit boards, clocks, flashlight, calculators, phones, etc. E-waste recycling is the reuse and reprocessing of electronics and electricals. This chapter discusses the subject of E-waste recycling in detail.

ELECTRONIC WASTE

Electronic waste, also called e-waste, are various forms of electric and electronic equipment that have ceased to be of value to their users or no longer satisfy their original purpose. Electronic waste (e-waste) products have exhausted their utility value through either redundancy, replacement, or breakage and include both "white goods" such as refrigerators, washing machines, and microwaves and "brown goods" such as televisions, radios, computers, and cell phones. Given that the information and technology revolution has exponentially increased the use of new electronic equipment, it has also produced growing volumes of obsolete products; e-waste is one of the fastest-growing waste streams. Although e-waste contains complex combinations of highly toxic substances that pose a danger to health and the environment, many of the products also contain recoverable precious materials, making it a different kind of waste compared with traditional municipal waste.

Globally, e-waste constitutes more than 5 percent of all municipal solid waste and is increasing with the rise of sales of electronic products in developing countries. The majority of the world's e-waste is recycled in developing countries, where informal and hazardous setups for the extraction and sale of metals are common. Recycling companies in developed countries face strict environmental regulatory regimes and an increasing cost of waste disposal and thus may find exportation to small traders in developing countries more profitable than recycling in their own countries. There is also significant illegal transboundary movement of e-waste in the form of donations and charity from rich industrialized nations to developing countries. E-waste profiteers can harvest substantial profits owing to lax environmental laws, corrupt officials, and poorly paid workers, and there is an urgent need to develop policies and strategies to dispose of and recycle e-waste safely in order to achieve a sustainable future.

Electronic waste Electronic waste in a garbage dump.

Impacts on Human Health

The complex composition and improper handling of e-waste adversely affect human health. A growing body of epidemiological and clinical evidence has led to increased concern about the potential threat of e-waste to human health, especially in developing countries such as India and China. The primitive methods used by unregulated backyard operators (e.g., the informal sector) to reclaim, reprocess, and recycle e-waste materials expose the workers to a number of toxic substances. Processes such as dismantling components, wet chemical processing, and incineration are used and result in direct exposure and inhalation of harmful chemicals. Safety equipment such as gloves, face masks, and ventilation fans are virtually unknown, and workers often have little idea of what they are handling.

For instance, in terms of health hazards, open burning of printed wiring boards increases the concentration of dioxins in the surrounding areas. These toxins cause an increased risk of cancer if inhaled by workers and local residents. Toxic metals and poison can also enter the bloodstream during the manual extraction and collection of tiny quantities of precious metals, and workers are continuously exposed to poisonous chemicals and fumes of highly concentrated acids. Recovering resalable copper by burning insulated wires causes neurological disorders, and acute exposure to cadmium, found in semiconductors and chip resistors, can damage the kidneys and liver and cause bone loss. Long-term exposure to lead on printed circuit boards and computer and television screens can damage the central and peripheral nervous system and kidneys, and children are more susceptible to these harmful effects.

Environmental Impacts

Although electronics constitute an indispensable part of everyday life, their hazardous effects on the environment cannot be overlooked or underestimated. The interface between electrical and electronic equipment and the environment takes place during the manufacturing, reprocessing, and disposal of these products. The emission of fumes, gases, and particulate matter into the air, the discharge of liquid waste into water and drainage systems, and the disposal of hazardous wastes contribute to environmental degradation. In addition to tighter regulation of e-waste recycling and disposal, there is a need for policies that extend the responsibility of all stakeholders, particularly the producers, beyond the point of sale and up to the end of product life.

There are a number of specific ways in which e-waste recycling can be damaging to the environment. Burning to recover metal from wires and cables leads to emissions of brominated and chlorinated dioxins, causing air pollution. During the recycling process in the informal sector, toxic chemicals that have no economic value are simply dumped. The toxic industrial effluent is poured into underground aquifers and seriously affects the local groundwater quality, thereby making the water unfit for human consumption or for agricultural purposes. Atmospheric pollution is caused by dismantling activities as dust particles loaded with heavy metals and flame retardants enter the atmosphere. These particles either redeposit (wet or dry deposition) near the emission source or, depending on their size, can be transported over long distances. The dust can also enter the soil or water systems and, with compounds found in wet and dry depositions, can leach into the ground and cause both soil and water pollution. Soils become toxic when substances such as lead, mercury, cadmium, arsenic, and polychlorinated biphenyls (PCBs) are deposited in landfills.

Classification

E-waste can be classified on the basis of its composition and components. Ferrous and nonferrous metals, glass, plastics, pollutants, and other are the six categories of materials reported for e-waste composition. Iron and steel constitute the major fraction in waste electrical and electronic equipment (WEEE) materials, with plastics being the second largest. Nonferrous materials, including metals such as copper and aluminum, and precious metals such as silver, gold, and platinum are third in abundance and have significant commercial value. Toxic materials include lead and cadmium in circuit boards, lead oxide and cadmium in cathode ray tubes, mercury in switches and flat-screen monitors, brominated flame retardants on printed circuit boards, and plastic and insulated cables; when these exceed the threshold quantities, they are regarded as pollutants and can damage the environment if disposed of improperly.

One of the most widely accepted classifications is based on European Union directives that divide e-waste into the 10 following categories:

- Large household appliances: Refrigerators, freezers, washing machines, clothes dryers, dishwashers, electric cooking stoves and hot plates, microwaves, electric fans, and air conditioners.

- Small household appliances: Vacuum cleaners, toasters, grinders, coffee machines, appliances for haircutting and drying, toothbrushing, and shaving.

- Information technology (IT) and telecommunications equipment: Mainframes, minicomputers, personal computers, laptops, notebooks, printers, telephones, and cell phones.

- Consumer equipment: Radios, televisions, video cameras, video recorders, stereo recorders, audio amplifiers, and musical instruments.

- Lighting equipment: Straight and compact fluorescent lamps and high-intensity discharge lamps.

- Electrical and electronic tools: Drills, saws, sewing machines, soldering irons, equipment for turning, milling, grinding, drilling, making holes, folding, bending, or similar processing of wood and metal.

- Toys, leisure equipment, and sporting goods: Electric trains or racing car sets, video games, and sports equipment with electric elements.

- Medical devices: Radiotherapy equipment, cardiology, dialysis, pulmonary ventilators, nuclear medicines, and analyzers.

- Monitoring and control instruments: Smoke detectors, heating regulators, and thermostats.

- Automatic dispensers: For hot drinks, hot or cold bottles, solid products, money, and all appliances that automatically deliver various products.

E-WASTE RECYCLING

Computer monitors are typically packed into low stacks on
wooden pallets for recycling and then shrink-wrapped.

Computer recycling, electronic recycling or e-waste recycling is the disassembly and separation of components and raw materials of waste electronics. Although the procedures of re-use, donation and repair are not strictly recycling, these are other common sustainable ways to dispose of IT waste.

In 2009, 38% of computers and a quarter of total electronic waste was recycled in the United States, 5% and 3% up from 3 years prior respectively. Since its inception in the early 1990s, more and more devices are recycled worldwide due to increased awareness and investment. Electronic recycling occurs primarily in order to recover valuable rare earth metals and precious metals, which are in short supply, as well as plastics and metals. These are resold or used in new devices after purification, in effect creating a circular economy. Such processes involve specialised facilities and premises, but within the home or ordinary workplace, sound components of damaged or obsolete computers can often be reused, reducing replacement costs.

Recycling is considered environmentally friendly because it prevents hazardous waste, including heavy metals and carcinogens, from entering the atmosphere, landfill or waterways. While electronics consist a small fraction of total waste generated, they are far more dangerous. There is stringent legislation designed to enforce and encourage the sustainable disposal of appliances, the most notable being the Waste Electrical and Electronic Equipment Directive of the European Union and the United States National Computer Recycling Act.

Reasons for Recycling

Obsolete computers and old electronics are valuable sources for secondary raw materials if recycled; otherwise, these devices are a source of toxins and carcinogens. Rapid technology change, low initial cost, and planned obsolescence have resulted in a fast-growing surplus of computers and other electronic components around the globe. Technical solutions are available, but in most cases a legal framework, collection system, logistics, and other services need to be implemented before applying a technical solution. The U.S. Environmental Protection Agency, estimates 30 to 40 million surplus PCs, classified as "hazardous household waste", would be ready for end-of-life management in the next few years. The U.S. National Safety Council estimates that 75% of all personal computers ever sold are now surplus electronics.

In 2007, the United States Environmental Protection Agency (EPA) stated that more than 63 million computers in the U.S. were traded in for replacements or discarded. Today, 15% of electronic devices and equipment are recycled in the United States. Most electronic waste is sent to landfills or incinerated, which releases materials such as lead, mercury, or cadmium into the soil, groundwater, and atmosphere, thus having a negative impact on the environment.

Many materials used in computer hardware can be recovered by recycling for use in future production. Reuse of tin, silicon, iron, aluminium, and a variety of plastics that are present in bulk in computers or other electronics can reduce the costs of constructing new systems. Components frequently contain copper, gold, tantalum, silver, platinum, palladium, and lead as well as other valuable materials suitable for reclamation.

Computer components contain many toxic substances, like dioxins, polychlorinated biphenyls (PCBs), cadmium, chromium, radioactive isotopes and mercury. A typical computer monitor may contain more than 6% lead by weight, much of which is in the lead glass of the cathode ray tube (CRT). A typical 15 inch (38 cm) computer monitor may contain 1.5 pounds (1 kg) of lead but other monitors have been estimated to have up to 8 pounds (4 kg) of lead. Circuit boards contain considerable quantities of lead-tin solders that are more likely to leach into groundwater or create air pollution due to incineration. In US landfills, about 40% of the lead content levels are from e-waste. The processing (e.g. incineration and acid treatments) required to reclaim these precious substances may release, generate, or synthesize toxic byproducts.

Export of waste to countries with lower environmental standards is a major concern. The Basel Convention includes hazardous wastes such as, but not limited to, CRT screens as an item that may not be exported transcontinentally without prior consent of both the country exporting and receiving the waste. Companies may find it cost-effective in the short term to sell outdated computers to less developed countries with lax regulations. It is commonly believed that a majority of surplus laptops are routed to developing nations. The high value of working and reusable laptops, computers, and components (e.g. RAM) can help pay the cost of transportation for many worthless commodities. Laws governing the exportation of waste electronics are put in place to govern recycling companies in developed countries which ship waste to Third World countries. However, concerns about the impact of e-recycling on human health, the health of recycling workers and environmental degradation remain. For example, due to the lack of strict regulations in developing countries, sometimes workers smash old products, propelling toxins on to the ground, contaminating the soil and putting those who do not wear shoes in danger. Other procedures include burning away wire insulation and acid baths to resell circuit boards. These methods pose environmental and health hazards, as toxins are released into the air and acid bath residue can enter the water supply.

Recycling Methods

Computers being collected for recycling at a pickup
event in Olympia, Washington, United States.

Consumer Recycling

Consumer recycling options consists of sale, donating computers directly to organizations in need, sending devices directly back to their original manufacturers, or getting components to a convenient recycler or refurbisher.

Scrapping/Recycling

In the recycling process, TVs, monitors, mobile phones and computers are typically tested for reuse and repaired. If broken, they may be disassembled for parts still having high value if labour is cheap enough. Other e-waste is shredded to roughly 100 mm pieces and manually checked to separate out toxic batteries and capacitors which contain poisonous metals. The remaining pieces are further shredded to ~10 mm and passed under a magnet to remove ferrous metals. An eddy current ejects non-ferrous metals, which are sorted by density either by a centrifuge or vibrating plates. Precious metals can be dissolved in acid, sorted, and smelted into ingots. The remaining glass and plastic fractions are separated by density and sold to re-processors. TVs and monitors must be manually disassembled to remove either toxic lead in CRTs or the mercury in flat screens.

Corporations face risks both for incompletely destroyed data and for improperly disposed computers. In the UK, some recycling companies use a specialized WEEE-registered contractor to dispose IT equipment and electrical appliances, who disposes it safely and legally. In America, companies are liable for compliance with regulations even if the recycling process is outsourced under the Resource Conservation and Recovery Act. Companies can mitigate these risks by requiring waivers of liability, audit trails, certificates of data destruction, signed confidentiality agreements, and random audits of information security. The National Association of Information Destruction is an international trade association for data destruction providers.

Sale

Online auctions are an alternative for consumers willing to resell for cash less fees, in a complicated, self-managed, competitive environment where paid listings might not sell. Online classified ads can be similarly risky due to forgery scams and uncertainty.

Take Back

When researching computer companies before a computer purchase, consumers can find out if they offer recycling services. Most major computer manufacturers offer some form of recycling. At the user's request they may mail in their old computers, or arrange for pickup from the manufacturer.

Hewlett-Packard also offers free recycling, but only one of its "national" recycling programs is available nationally, rather than in one or two specific states. Hewlett-Packard also offers to pick up any computer product of any brand for a fee, and to offer a coupon against the purchase of future computers or components; it was the largest computer recycler in America in 2003, and it has recycled over 750,000,000 pounds (340,000,000 kg) of electronic waste globally since 1995. It encourages the shared approach of collection points for consumers and recyclers to meet.

Exchange

Manufacturers often offer a free replacement service when purchasing a new PC. Dell Computers and Apple Inc. take back old products when one buys a new one. Both refurbish and resell their own computers with a one-year warranty.

Many companies purchase and recycle all brands of working and broken laptops and notebook computers from individuals and corporations. Building a market for recycling of desktop computers has proven more difficult than exchange programs for laptops, smartphones and other smaller electronics. A basic business model is to provide a seller an instant online quote based on laptop characteristics, then to send a shipping label and prepaid box to the seller, to erase, reformat, and process the laptop, and to pay rapidly by cheque. A majority of these companies are also generalized electronic waste recyclers as well; organizations that recycle computers exclusively include Cash For Laptops, a laptop refurbisher in Nevada that claims to be the first to buy laptops online, in 2001.

Donations/Nonprofits

With the constant rising costs due to inflation, many families or schools do not have the sufficient funds available for computers to be utilized along with education standards. Families also impacted by disaster suffer as well due to the financial impact of the situation they have incurred. Many nonprofit organizations, such as InterConnection.org, can be found locally as well as around the web and give detailed descriptions as to what methods are used for dissemination and detailed instructions on how to donate. The impact can be seen locally and globally, affecting thousands of those in need. In Canada non profit organizations engaged in computer recycling, such as The Electronic Recycling Association Calgary, Edmonton, Vancouver, Winnipeg, Toronto, Montreal, Computers for Schools Canada wide, are very active in collecting and refurbishing computers and laptops to help the non profit and charitable sectors and schools.

Junkyard Computing

The term *junkyard computing* is a colloquial expression for using old or inferior hardware to fulfill computational tasks while handling reliability and availability on software level. It utilizes abstraction of computational resources via software, allowing hardware replacement at very low effort. Ease of replacement is hereby a corner point since hardware failures are expected at any time due to the condition of the underlying infrastructure. This paradigm became more widely used with the introduction of cluster orchestration software like Kubernetes or Apache Mesos, since large monolithic applications require reliability and availability on machine level whereas this kind of software is fault tolerant by design. Those orchestration tools also introduced fairly fast set-up processes allowing to use junkyard computing economically and even making this pattern applicable in the first place. Further use cases were introduced when continuous delivery was getting more widely accepted. Infrastructure to execute tests and static code analysis was needed which requires as much performance as possible while being extremely cost effective. From an economical and technological perspective, junkyard computing is only practicable for a small number of users or companies. It already requires a decent number of physical machines to compensate for hardware failures while maintaining the required reliability and availability. This implies a direct need for a matching underling infrastructure to house all the computers and servers. Scaling this paradigm is also quite limited due to the increasing importance of factors like power efficiency and maintenance efforts, making this kind of computing perfect for mid-sized applications.

Data Security

Electronic waste dump at Agbogbloshie, Ghana. Organized criminals
commonly search the drives for information to use in local scams.

E-waste presents a potential security threat to individuals and exporting countries. Hard drives that are not properly erased before the computer is disposed of can be reopened, exposing sensitive information. Credit card numbers, private financial data, account information and records of online transactions can be accessed by most willing individuals. Organized criminals in Ghana commonly search the drives for information to use in local scams.

Government contracts have been discovered on hard drives found in Agbogbloshie, Ghana. Multimillion-dollar agreements from United States security institutions such as the Defense Intelligence Agency (DIA), the Transportation Security Administration and Homeland Security have all resurfaced in Agbogbloshie.

Reasons to Destroy and Recycle Securely

There are ways to ensure that not only hardware is destroyed but also the private data on the hard drive. Having customer data stolen, lost, or misplaced contributes to the ever-growing number of people who are affected by identity theft, which can cause corporations to lose more than just money. The image of a company that holds secure data, such as banks, law firms, pharmaceuticals, and credit corporations is also at risk. If a company's public image is hurt, it could cause consumers to not use their services and could cost millions in business losses and positive public relation campaigns. The cost of data breaches "varies widely, ranging from $90 to $50,000 (under HIPAA's new HITECH amendment, that came about through the American Recovery and Revitalization act of 2009), as per customer record, depending on whether the breach is "low-profile" or "high-profile" and the company is in a non-regulated or highly regulated area, such as banking or medical institutions."

There is also a major backlash from the consumer if there is a data breach in a company that is supposed to be trusted to protect their private information. If an organization has any consumer info on file, they must by law (Red Flags Clarification act of 2010) have written information protection policies and procedures in place, that serve to combat, mitigate, and detect vulnerable areas that could result in identity theft. The United States Department of Defense has published a standard to which recyclers and individuals may meet in order to satisfy HIPAA requirements.

Secure Recycling

Countries have developed standards, aimed at businesses and with the purpose of ensuring the security of Data contained in 'confidential' computer media. National Association for Information Destruction (NAID) "is the international trade association for companies providing information destruction services. Suppliers of products, equipment and services to destruction companies are also eligible for membership. NAID's mission is to promote the information destruction industry and the standards and ethics of its member companies." There are companies that follow the guidelines from NAID and also meet all Federal EPA and local DEP regulations. The typical process for computer recycling aims to securely destroy hard drives while still recycling the byproduct. A typical process for effective computer recycling:

- Receive hardware for destruction in locked and securely transported vehicles.
- Shred hard drives.
- Separate all aluminum from the waste metals with an electromagnet.
- Collect and securely deliver the shredded remains to an aluminum recycling plant.
- Mold the remaining hard drive parts into aluminum ingots.

The Asset Disposal and Information Security Alliance (ADISA) publishes an *ADISA IT Asset Disposal Security Standard* that covers all phases of the e-waste disposal process from collection to transportation, storage and sanitization's at the disposal facility. It also conducts periodic audits of disposal vendors.

GENERATION, COMPOSITION, COLLECTION, TREATMENT AND DISPOSAL SYSTEM

The information technology (IT) industry is an important engine of growth of any country. With the rapid development of technology, manufacturers now produce superior televisions, new and smarter mobile phones, and new computing devices at an increasing rate. People are enjoying what technology brings, surfing the Internet on their smart phones or tablets and watching high-definition movies on their televisions at home. As more and more electronic products are produced to fulfill the needs of people worldwide, more resources are used to produce these items. Hence, the rapid growth of computing and other information and communication equipment is driving the ever-increasing production of electronic waste (e-waste). The current e-waste encompasses a particularly complex waste flow in terms of the variety of products. Over the next few years, one billion computers will be obsolete. In 2005, 8.3-9.1 million tons of e-waste was produced across the 27 members of the European Union (EU). By 2020, the total waste electrical and electronic equipment (WEEE) is estimated to grow between 2.5% and 2.7% annually, reaching a total of approximately 12.3 million tons. The reason is that the number of appliances entering the market every year is increasing in developed and developing countries. Sales of electronic products in countries such as China and India and across Africa and Latin America are predicted to rise sharply in the next 10 years.

Also, it is a higher growth pattern that will be influenced not only by need but also by changes in technology, design, and marketing. The diverse waste generated due to advancement of technology may have significant impacts on the environment and public, if not properly stored, collected, transported, treated, and disposed of. Thus, around the globe, e-waste generation, treatment, and disposal are becoming issues of concern to waste management professionals, innumerable non-governmental organizations and citizens, and international agencies and governments, particularly in developing and transition countries. E-waste stream contains diverse materials, which requires special treatment and cannot be dumped in landfill sites, most prominently, hazardous substances such as lead, polychlorinated biphenyls (PCBs), polybrominated biphenyls (PBBs), mercury, polybrominated diphenyl ethers (PBDEs), brominated flame retardants (BFRs), and valuable substances such as iron, steel, copper, aluminium, gold, silver, platinum, palladium, and plastics.

During the last decade, large amounts of diverse e-waste discarded by developing and transition countries, as well as a sizeable portion of the e-waste generated from developed countries and exported to developing and transition countries, has been rapidly piling up in developing countries impacting their emerging economies. The management of e-waste in developing and transition countries is exacerbated by several factors, including illegal trafficking and unlicensed recycling of e-waste; lack of technological, financial, and technically skilled human resources; inadequate organizational structure required; and an inadequate description of the roles and responsibilities of stakeholders and institutions involved in e-waste management. In Africa, e-waste management is still in its infancy; characterized by little available information on the e-waste situation, the recovery of valuable materials in small workshops using rudimentary recycling methods, lack of awareness on the impacts of e-waste, and the total absence of policy specifically dealing with e-waste.

E-waste Generation

The major problem associated with e-waste management is its ever increasing quantum. However, the e-waste quantities represent a small percentage of the overall municipal solid waste (MSW). Data on e-waste generation may vary between areas of a country because of the definitions of waste arising, technological equipment used, the consumption patterns of the consumers, and changes in the living standards across the globe. Global e-waste generated per year amounts to approximately 20-25 million tons, most of which is being produced in rich nations such as the United States (US) or European Union member countries. The US, is the largest generator of e-waste, with a total accumulation of 3 million tons per year; and China is the second largest, producing 2.3 million tons each year. Brazil generates the second greatest quantity of e-waste among emerging countries.

In Malaysia, the volume of e-waste generated is estimated at roughly 0.8-1.3 kg of waste per capita per day, with an increasing trend of e-waste generation, which rose to 134,000 tons in 2009. Furthermore, the volume of e-waste in Malaysia is expected to rise to 1.1 million metric tons in 2020, at an annual rate of 14%. In South Africa and China, e-waste production from old computers will increase by 200-400% from 2007 to 2020, and by 500% . In this same period e-waste from televisions will be 1.5-2 times higher in China and India; whereas , e-waste from discarded refrigerators will double or triple by 2020. These data only include e-waste generated nationally and do not include waste imports (both legal and illegal) which are substantial in emerging economies such as India and China. The reason is that large amount of WEEE enters India from foreign countries without paying any duty in the name of charity. The rate at which the e-waste volume is increasing globally is 5 to 10% yearly.

Composition of E-waste

E-waste normally contains valuable, as well as potentially toxic materials. The composition of e-waste depends strongly on factors such as the type of electronic device, the model, manufacturer, date of manufacture, and the age of the scrap. Scrap from IT and telecommunication systems contain a higher amount of precious metals than scrap from household appliances. For instance, a mobile phone contains more than 40 elements, base metals such as copper (Cu) and tin (Sn); special metals such as lithium (Li) cobalt (Co), indium (In), and antimony (Sb); and precious metals such as silver (Ag), gold (Au), and palladium (Pd). Special treatment of e-waste should be considered to prevent wasting valuable materials and rare elements. Materials such as gold and palladium can be mined more effectively from e-waste compared to mining from ore.

By contrast, e-waste contains PBDEs, which are flame retardants that are mixed into plastics and other components. Circuit boards found in most of the electronic devices may contain arsenic (As), cadmium (Cd), chromium (Cr), lead (Pb), mercury (Hg), and other toxic chemicals. Typical printed circuit boards treated with lead solder in electronic devices contain approximately 50 g of tin-lead solder per square meter of circuit board. Obsolete refrigerators, freezers, and air conditioning units contain ozone depleting Chlorofluorocarbons (CFCs). The prominent materials such as barium, cadmium, copper, lead, zinc, and other rare earth metals are contained in end-of-life (EOL) cathode ray tubes (CRTs) in computer monitors, and televisions. For example, items such as leaded glass provide protection against X-rays produced in the picture projection process in CRTs. The average lead in CTR monitors is 1.6-3.2 kg.

Thus, the US and other developed countries in the EU and Japan have banned the disposal of cathode ray tubes in landfills because of their toxic characteristics. A critical challenge in designing and developing strategies to manage e-waste is the changing composition of the many constituents due the advancement of technology, particularly in the electronic components. It is against this background that e-waste recycling and disposal methods ought to keep pace with the changing composition of e-waste. Several factors influence the composition of e-waste, including economic conditions, availability of a reuse market, and infrastructure of the recycling industry, waste segregation programs, and regulation enforcement. Figure illustrates the distinctive materials in a WEEE.

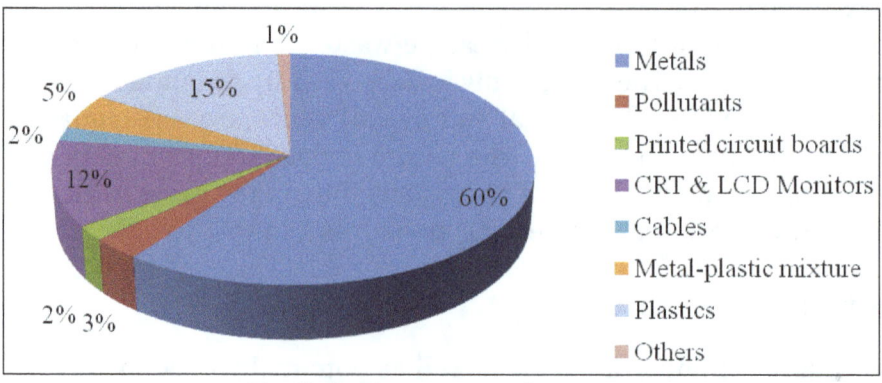

The distinctive contents of a WEEE.

E-waste Data for Several Countries across the Globe

Amount of E-waste Collected and Treated

E-waste generated from the different diverse sources is normally collected as a whole unit or sub-unit of functional equipment. In many instances across the globe, whole units of e-waste have been categorized as e-waste. Based on the number of discarded information communication technology (ICT) devices collected in Europe, computers, cell phones, fixed-line telephones, televisions, and radios are the major electronic products, and together they amounted to 11.7 million tons in 2007. In 2004, approximately 75,000 tons of WEEE were collected, classified, disassembled, and then processed in Switzerland, compared with the collection of approximately 68,000 tons in 2003.

In developing and transition countries, little consideration is given to the quantification of the e-waste collected. The reason is that in pre-reprocessing stages, collection of the e-waste is mostly undertaken by the unorganized sector of scrap dealers/traders or peddlers. As a result, this information is invisible to the statistics collection system, which makes quantification of e-waste very difficult in developing and transition countries. More precise figures regarding unused electronic and electrical equipment/ waste electronic and electrical equipment (UEEE/WEEE) are not available because the customs data do not distinguish between used and new equipment and the import statistics reveal only total values. Based on the current understanding on e-waste management, research studies suggest that to achieve sustainable development goals associated with waste management would require successful establishment of baseline levels of information from which more informed e-waste management and policy decisions can be made. Similarly, to effectively manage e-waste could require establishment of separate collection channels that would be environmentally friendly. This could result in the reduction of e-waste generated and its environmental impacts.

In the EU, the EU WEEE directive clearly imposes collection, recovery, and recycling targets on its member countries. Thus, it stipulates a minimum collection target of 4 kg/capita per year for all the member states. These collection- and weight-based recycling targets seek to reduce the amount of hazardous substances disposed into landfills and to increase the availability of recyclable materials that indirectly encourages less virgin materials consumption in new products. Switzerland is the first country in the world to have established and implemented a formal e-waste management system that has recycled 11 kg/capita of WEEE against the target of 4 kg/capita set by the EU. One-third of electrical and electronic waste in the EU is reported as separately collected and appropriately treated. In 2006, Germany collected and treated about 754,000 tons of e-waste according to the ElektroG system, while other EU member states collected about 19,000 tons. It was also forecasted that IT and telecommunications equipment put on the market were 315,000 tons, and the waste collected and treated in the system according to ElektroG was about 102,000 tons (7,000 tons of this was collected from other EU countries). This shows the effective collection and treatment of e-waste in the

EU. The introduction of the extended producer responsibility (EPR) scheme in 2003 was the most important step in South Korea, and about 70% of e-waste was collected by producers. Over the same period, the amount of e-waste reused and recycled was 12% and 69% respectively. The remainder was sent to landfill sites or incineration plants, accounting for 19%.

Amount of E-waste Disposed

The scientific and environment friendly disposal of e-waste is critical. Relevant past studies on e-waste management confirmed that rapid growth combined with rapid product obsolescence are the most important factors making discarded e-products one the fastest growing waste fraction, accounting for 8% of all municipal waste in the EU. If not disposed of properly it could lead to significant negative environmental impacts. The average for developing and transition countries was 1% of total solid waste, which increased to 2% in 2010. Developing and transition countries, especially those in Africa and Asia, are the primary destinations for e-waste dumping, despite these countries lacking basic disposal technologies or facilities.

In 2012, more than 70% of the total electronic waste collected worldwide was actually exported or discarded by developed countries. In the US alone, 130,000 computers and more than 300,000 cell phones are disposed each day, and an estimated 80% of the generated e-waste is sent to less-developed countries. In 2007 in the US, 410 thousand tons were recycled (13.6%), and the rest was improperly discharged in landfills or incinerated. Between 2003 and 2005, approximately 80-85% of the e-waste ready for EOL management ended up in US landfills. A related study about e-waste management in the US pointed out that in 2009, enormous quantities of e-waste (82.3%) was disposed in landfill sites and incinerators, while 17.7% went to the recyclers. In the EU, it is shown that two-thirds of this waste stream is potentially still going to landfills and to sub-standard treatment sites in or outside the EU. In China, huge volumes of e-waste have been discarded in recent years as people more frequently replace their old home appliances with new ones.

A relevant case-study on e-waste management pointed out that it is not possible to make an overall comparison between different countries, even if they are in the same continent, as the definitions in legislation and categorization of e-waste streams differ. Nevertheless, it is established that the main volumes of e-waste reside in developed countries.

Collection, Treatment and Disposal Systems

Collection, treatment, and disposal systems are critical elements of e-waste management. Most developed countries have framed conventions, directives, and laws aimed at fostering proper collection, treatment, and recycling of e-waste, as well as safe disposal of the non-recyclable components. These include the EPR, product stewardship, advance recycling fund (ARF), the 3Rs or Reduce, Reuse, Recycle initiative, etc. For the EU, two directives have been promulgated to place an obligation on the producers of e-goods to take back EOL or waste products free of charge in an effort to reduce the amount of waste going to landfills. However, in developing and transition countries, e-waste is treated in backyard operations, using open sky incineration, cyanide leaching, and simple smelters to recover precious metals mainly copper, gold, and silver—with comparatively low yields—and discarding the rest with municipal solid waste at open dumps, into surface water bodies and at unlined and unmonitored landfills, thereby causing adverse environmental and health effects. Table presents a comparison of typical e-waste treatment processes in developed and developing countries.

Table: Comparison of typical e-waste treatment processes in developed and developing countries.

Developing countries	Developed countries
"Informal" sector	Formal sector
Manual dismantling	Manual dismantling
Manual separation	Semi-automation separation
Recovery of metals by heating, burning, and acid leaching of e-waste scrap in small workshops	Recovery of metals by the state-of-the-art methods in smelter and refineries

Disposal System

Disposal of e-waste is mainly through landfilling. Most often, the discarded electronic goods finally end-up in landfill sites along with other municipal waste or are openly burnt releasing toxic and carcinogenic substances into the atmosphere. In developing and transition countries the disposal of e-waste in the informal sector is very rudimentary so far as the safe techniques employed and practices are concerned, resulting in low recovery of materials. Table presents a comparison of typical disposal systems in developed and developing countries.

Table: Comparison of a typical e-waste disposal systems in developed and developing countries.

Developed countries	Developing countries
Incineration with MSW	Opening burning
Landfill disposal	Open dumping

E-waste management is different between developed countries and developing and transition countries. Developing and transition countries do not have guidelines and information campaigns on the fate of e-waste. Especially, less sophisticated disposal systems are used, from open burning and dumping to uncontrolled landfill sites, which pose significant environmental pollution and occupational exposure to e-waste-derived chemicals. Serious challenges in the disposal of e-waste were analyzed across developing countries such as Brazil, China, and India, outlining the difficulty to implement/enforce existing regulations and clean technologies backed by lack of capacity building and awareness. In contrast, developed countries have devised sophisticated disposal schemes and high-cost systems, which are less hazardous to handle waste. However, a comprehensive overview of the situation is constrained by the availability of data. This means that the differences in the socio-economic and legal contexts between typical developing and developed countries' scenarios limit e-waste management in developing and transition countries. The regulations that guide the disposition of e-waste in developing countries is mostly fragmented and lack monitoring, while in developed countries the regulations are stringent and there is effective monitoring.

E-waste Collection Schemes in Different Parts of the World with a Regional Focus

In general, citizens must sort and segregate e-waste to divert e-waste from mixed municipal waste collection schemes and landfills due to the heteremeneous materials it contains. It needs to be stored, and then transferred to the curbside or transported to an offsite collection site. Although research has supported that curbside collection is the most convenient collection system for households, offsite drop-off remains attractive to waste management authorities. This is because curbside collections are regarded as expensive, time-consuming to design, implement, and operate.

In essence, a separate, parallel collection and management scheme is required, organized by the authorities, the producers, or retailers. Compared to simple or commingled collection, such as single-stream collection, source separation imposes additional efforts on citizens regarding material segregation and drop-off and, thus, convenience is of paramount importance. In developed countries, e-waste is collected to recover some materials of value and to be safely rid of the lead, cadmium, mercury, dioxins, furans, and such toxic materials they contain. On the other hand, in developing countries, e-waste is collected principally to recover a few metals of value. E-waste collection is logically a profit-driven activity. E-waste contains a huge volume of different engineering materials that can be reused via available and evolving technologies.

Asian Region

In Malaysia, a planned infrastructure is being promoted for whole units of WEEE to be collected from households, business entities, and institutions. The Department of Environment (DoE) and the Japanese International Cooperation (JICA) are trying to develop an e-waste collection model for household items in Penang state for the very first time. This model is expected to be used to make a countrywide drive after the model's test run, which may happen in the next few years. However, this model has limitations, and only can ensure the collection of a small portion of e-waste. Thus, there is no engineering analysis on material characteristics, remanufacturing potential, and economic benefits, and an optimization analysis is not yet planned. Moreover, there is no reverse logistic system in this model. The e-waste collection activities in Malaysia include: DoE-licensed contractors, retailer's collection, environmental working groups, voluntary collection organization, social organizations, informal scrap collectors, street buyers, scavengers, traditional hawkers (Surat khabar lama), and manufacturers' initiatives such as Panasonic Malaysia ECOMOTO Take back, Nokia Malaysia, Dell Malaysia HP, and Pikom (National ICT).

In other Asian countries, collection of most -waste materials and components remains in the hands of the informal sector. "Scavenging" or the informal sector is the predominant collection scheme of e-waste in the Asian region. Using inappropriate methods, this poses a severe threat to the environment and health of the workers. For instance, in China, Taiwan, Thailand, the Philippines, Indonesia, and other neighboring countries, this informal stream of e-waste collection is not under regulation, and most of the e-waste ends up in landfills through the informal stream. Furthermore, collection systems and procedures in the region are very loose, and there is limited established market for finished products resulting from recycling. Customers need to be given incentives to return their EOL e-products back to the collection centers. and China, studies equivocally state that consumers look for economic benefits for discarding their e-waste. Thus, the Chinese residents, in the likelihood of a take-back regime, reportedly seem to prefer the pay-in-advance scheme against the deposit-refund route favored by residents . There exists a very well networked and effective door-to-door collection network . China has established special recovery industrial parks in Tianjin, Taicang, Ningbo, Linyi, Liaozhong, Taizhou, and Zhangzhou in order to promote efficient and environmentally friendly recovery of original and imported metals. The collection of discarded household electronic and electrical equipment in China is still dominated by the so-called informal individual collectors (peddlers). They provide a door-to-door service by paying marginal fees to e-waste owners and then sell them to e-waste dealers.

European Union Context

Consumers in Europe use municipal collection, retailer collection, social organization collection, and the re-use market to collect e-waste. The so-called municipal collection is performed by local authorities (municipalities or counties). It is pointed out that some municipalities collect the WEEE themselves, while others themselves, while others contract with other parties to collect to collect it on their behalf. Municipal collection activities are managed and financed by public waste management entities, whereby drop-off points and doorstep collection are used. Retailer collection is performed either by the retailers themselves or by their logistics partners who deliver new appliances to consumers. Social organization collection is performed in cooperation with several members of the reverse supply chain, with the purpose of providing a material input to and a financial benefit for the social organizations. The re-use market extends the use phase of appliances, thereby delaying the final discarding by the ultimate owner/user of the appliance into municipal, retailer, or social collection. Germany has developed a curbside collection scheme and is already achieving remarkable success in e-waste management andrecycling. The typical collection channels in the EU, from dismantling through pre-processing until end-processing, lead to the safe disposal or processing of e-waste.

Situation in the US and Canada

The US and Canadian provinces are increasingly adopting EPR and product steward-ship (PS) schemes for WEEE. For instance, in the state of Maine in the US, the WEEE management program is based on a PS scheme, with the active participation of retail-ers. Three American-based non-governmental organizations (NGOs) are particularly active in e-waste issues. The Basel Action Network (BAN), Silicon Valley Toxic Coa-lition (SVTC), and Electronics Take-Back Coalition (ETBC) constitute an associated network of environmental advocacy NGOs in the US. The three organizations' common objective is to promote national-level solutions for hazardous waste management. A recent initiative has been e-Stewards, a system for auditing and certifying recyclers and take-back programs so that conscientious consumers know which ones meet high standards. Canada is among the countries developing systems based on these princi-ples and EPR. Also, Canada has well-developed and advanced collection systems. In the US, Apple, Sony, Sharp, Mitsubishi, Samsung, Hewlett-Packard, Dell, LG, Lenovo, Panasonic, and Toshiba have free collection point or mail-in take-back programs of their products.

Illegal E-waste Trade and Illegal Waste Disposal Practices

Across the globe, high volumes of e-waste have been discarded in recent years. Despite the fact that many countries have already organized e-waste regulations, there are ad-ditional problems with the import/export of e-waste. For instance, in industrialized countries such as the US, Japan, and the EU, recycling operations have set high envi-ronmental and social standards, which trigger the illegal exportation of WEEE to de-veloping and transition countries. The developing and transition countries lack cleaner technologies, waste minimization measures, and environmental sound management systems. As a result, the items are treated, recycled and reused with less consideration for environmental protection and public safety and health.

Several countries have ratified the Basel Convention on trans-boundary movement of hazardous waste. It specifies the relevant requirements of governments exporting hazardous waste, and stipulates the responsibility of the government of the importing country. However, because of the lack of management systems for secondhand e-prod-ucts and e-scraps, these items are not covered by the convention's rules. The Basel Convention does not solve the new environmental problems caused by the recycling of e-waste. Over the recent years, the exportation of secondhand electronic devices from developed countries to developing and transition countries continues through clandestine operations, legal loopholes, and by countries that have not ratified the convention. For instance, about 2 million secondhand televisions, approximately 400,000 units are exported from Japan to the Philippines, annually. However, inappropriate recycling and final treatment processes such as open burning of wires and improper crushing of CRT tubes has been observed at or near dumpsites in Manila. Amendments to the Basel Convention are necessary to prevent the exportation of hazardous from developed countries to developing and transition countries for any purpose (even for recycling).

China, Vietnam, and Cambodia have built up their own legal frameworks to deal with the import of secondhand items and hazardous wastes. For instance, in 1996, Cambodia banned the importation computers because of concerns about the possibility of spreading virus infections into domestic computer systems. Nevertheless, e-waste scrap is not subjected to any legal regulations.

In 2000, China introduced a complete ban on the importation of secondhand EEE. It also prohibited the importation of printed circuit boards. In 2001, Vietnam followed suit to introduce the ban on importation of secondhand EEE, including home appliances and computers. Between 2004 and 2006, Vietnam introduced laws to tighten the ban on the importation of secondhand EEE and re-exportation of e-waste scrap by the Minister for Trade. Along with laws banning the importation of secondhand EEE, relevant prohibitions on the importation of e-waste scrap for any purpose and on the dismantling of e-waste scrap have been enacted in July 2005. Although bans on the importation of secondhand EEE and printed circuit boards have been introduced in China and Vietnam, research studies pointed out that due to the demand for used electronic products and used parts, significant proportions of these materials still find their way into these two countries. In addition, these countries lack effective implementation of policies and monitoring measures. For instance, China allows the importation of secondhand EEEs to be imported as long as they are built and then re-exported. It is predicted that annually, some 57,700 tons of e-wastes were illegally imported, of which 8,470 tons were exported again. Also, mandatory removal results in spreading of improper recycling activities to other places. Given this background, it is clear that a major portion of e-waste scrap, such as printed circuit boards, has been, and is being, recycled or smuggled into Vietnam, China, and Cambodia.

The illegal trade of electrical and electronic waste to non-EU countries continues to be uncovered at EU borders. Past research studies confirmed that significant proportions of materials are still exported illegally outside of the EU member states because recycling companies, scrap dealers, brokers, and the so-called re-use companies take advantage of low dumping costs and environmental standards. Illegal dumping remains a serious problem in Japan, and some e-waste is exported overseas as reusable parts. China, along with Peru, Ghana, Nigeria, India, and Pakistan are the biggest recipients of e-waste from industrialized countries. Other leading recipient countries of e-waste are Singapore, Malaysia, Vietnam, Philippines, and Indonesia.

Approximately 500 containers with electrical and electronic equipment reach Nigeria every month. Some researchers estimate that approximately 400,000 used computers are imported every month. Of these, only approximately 50% are functional. Approximately 45% of the equipment comes from Europe and the USA each, and the other 10% from Asia. This situation was also found in Ghana, where computers, televisions, and monitors were the most common imports. According to the available data, around 300 containers of UEEE/WEEE reach Ghana every month through the ports of Tema. The highest number of equipment from the EU comes from Germany, the Netherlands, and the UK. It was established that approximately 75-80% of the imported UEEE/WEEE cannot be reused.

An Outline of the Illegal Waste Disposal Practices associated with E-waste Fraction

In developing and transition countries, formal recycling of e-waste using efficient technologies and facilities is rare; therefore, e-waste is managed through various low-end management alternatives, such as disposal in open dumps, backyard recycling, and disposal into the environment, such as surface water, conventional landfills, etc. The majority of the unusable components are thrown away arbitrarily, polluting the environment and water sources. Developing and transitional countries have not yet established official e-waste recycling facilities. Some developing countries, such as South Africa, Indonesia, India, etc., have industrial areas where recycling facilities and plants have been built. However, backyard recycling of PCs, television sets, etc. is a common practice. For instance, individuals from the informal sector usually recover precious materials from e-waste, such as gold from the integrated circuit (IC) socket or IC chipset. Using their bare hands and without wearing any personal protective clothing (PPP) for safety and health protection mask, they burn ICs and mix the residue with other chemicals (e.g., nitric acid (HNO_3), selenium, etc.) to recover gold. This process generates waste water containing heavy metals that exceed World Health Organization (WHO) threshold values of waste water regulations (e.g., Cu, Cr, Co, Pb, nickel (Ni), Sn, and zinc (Zn)).

References

- Bakhiyi, Bouchra; Gravel, Sabrina; Ceballos, Diana; Flynn, Michael A.; Zayed, Joseph (2018). "Has the question of e-waste opened a Pandora's box? An overview of unpredictable issues and challenges". Environment International. 110: 173–192. doi:10.1016/j.envint.2017.10.021. PMID 29122313

- The-generation-composition-collection-treatment-and-disposal-system-and-impact-of-e-waste, e-waste-in-transition-from-pollution-to-resource: intechopen.com, Retrieved 19 March, 2020

- Judkis, Maura (2008-07-30). "4 Ways to Earn Cash for Recycling". Fresh Greens. U.S. News and World Report. Retrieved 2008-03-05

- A-review-of-technology-of-metal-recovery-from-electronic-waste, e-waste-in-transition-from-pollution-to-resource: intechopen.com, Retrieved 28 August, 2020

- Nguemaleu, Raoul-Abelin Choumin; Montheu, Lionel (2014-05-09). Roadmap to Greener Computing. CRC Press. p. 170. ISBN 9781466506848

5

Vehicle Recycling

Vehicles, at the end of their useful life, are dismantled for their spare parts and the process is termed as vehicle recycling. Different kinds of vehicles can be recycled such as cars, ships, trucks, etc. This chapter closely examines vehicle recycling to provide an extensive understanding of the subject.

Vehicle recycling is the dismantling of vehicles for spare parts. Motor vehicles are the premier recycled consumer product largely due to the work performed by auto recyclers at motor vehicle salvage yards (MVSYs). Over four million end-of-life vehicles (ELVs) are recycled annually in the United States. Auto recyclers dismantle ELVs to recover fluids and parts for reuse, and scrap material for recycling. Typically, auto recyclers manage to reuse and recycle over 75 percent of the material content of a vehicle, by weight. This trumps the recycling rate for aluminum cans the next most recycled consumer product at 61 percent.

The auto recycling business is over 75 years old and has evolved into a sophisticated market and technology-driven industry that must constantly change in response to innovations in automotive technology and manufacturing techniques. To be competitive and profitable in today's markets, the auto recycling process must involve much more than merely crushing wrecked, abandoned, and worn-out motor vehicles. The modern-day auto recycler must establish operating practices that realize the maximum market value of every ELV taken in and produce real economic and environmental benefits within the community being served.

Upon arrival at a well-run modern MVSY, inoperative motor vehicles are inspected for leaks and temporarily stored until further processing, as follows:

Motor vehicle fluids are drained separately over an impervious surface and put into leak-proof containers for reuse. For example: recovered gasoline is usually used to fuel company and employee vehicles. Recovered oil is used to fuel special furnaces that heat the facility. And, antifreeze is recycled and made available for reuse by customers.

Reusable parts are removed from the vehicle, cleaned, tested, inventoried, and stored in a warehouse for sale to auto repair garages, auto body shops, and individuals. Selected items include engines, transmissions, auto body parts, tires, radios, and other components.

After the vehicle is stripped of reusable parts, the remaining "hulk" is then stored until market conditions are favorable for sending it to a scrap processor for material recycling.

Most auto recyclers hire contractors to crush their vehicles using mobile auto crushing units. However, some auto recyclers have their own stationary crushing units. To properly prepare a vehicle for crushing, after all fluids should be drained; the fuel tank and radiator should be removed; the Freon or other air conditioning system refrigerants should be evacuated using certified equipment; and other recyclable and potentially hazardous components should be removed, including the battery, tires, air bag cartridges, mercury-containing switches, lead parts, and catalytic converters.

When market conditions are favorable, the crushed hulks are sent to a scrap processor, where the hulks are fed into a shredder for recovery of ferrous and non-ferrous metal for recycling. The remaining material, comprised mainly of plastic, foam, glass, textiles, and other materials (known as "auto shredder residue") is usually land filled.

Many professional auto recyclers use computer and satellite communication systems that enable direct inventory assessment and location of parts at other MVSYs on the network.

CAR RECYCLING

Sometimes when a vehicle reaches the end of its life the only thing to do is ship it off to a car recycler or, as the industry refers to them; Authorised Treatment Facilities (ATFs). You may choose to get your car recycled for a variety of reasons but regardless of whether it's due to a failed engine or MOT, crash damage or just old age, you've come to the right place.

What is the Process of Recycling a Car?

Once a vehicle has reached the end of its life it will be taken to an Authorised Treatment Facility where it will go through the following three stages:

- Depollution,
- Dismantlement,
- Destruction.

Depollution

When a vehicle goes through the depollution process it means that it is having all of the hazardous materials or components safely removed as required by the Environmental Agency. Depollution is important because end of life vehicles have various elements with the potential to harm the environment. Windscreen wash, coolants, antifreeze, batteries, oil and fuel are all examples of contaminants which need to disposed of with care to protect both the environment, and people, as if one of those substances were to enter the food chain or water supply, it would create an array of health problems. These substances must therefore be removed by professionals to ensure that they are disposed of properly.

Once a qualified person has depolluted a vehicle the parts will be stored and passed on to specialists who will break them down to reuse as something else. For example; tyres are often shredded and broken down into pellets before being reused as football pitch flooring. Whereas car batteries will be removed and either sold as spares, or disassembled for their components. Battery components include; distilled water, silver, lead, acid and plastic and once the ATF has neutralised the acid and purified the water for safe disposal, they will then melt down the metals to be repurposed. Only ATF's can do this as all hazardous substances must be recycled and disposed of by licensed specialists.

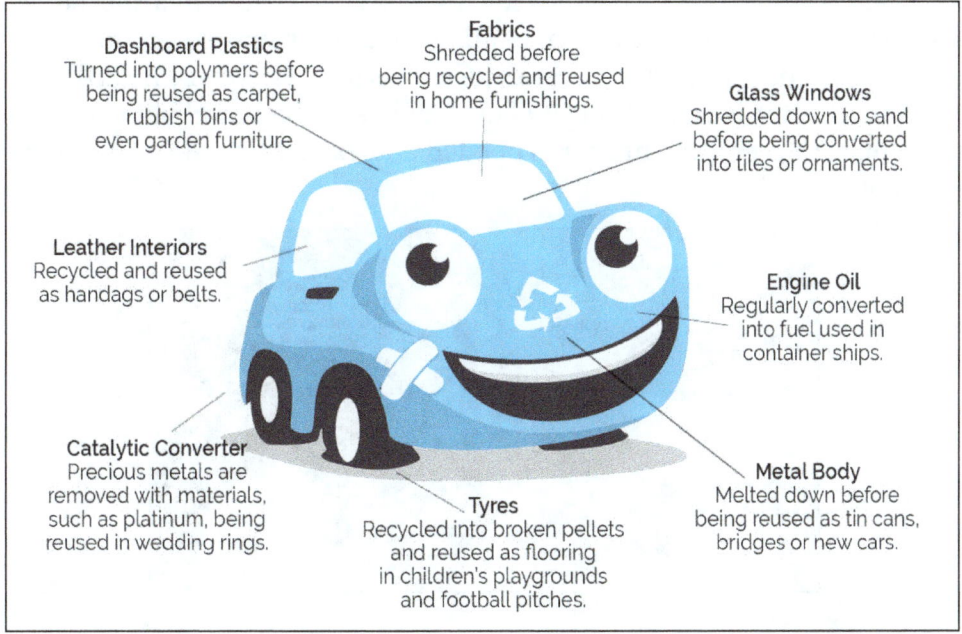

When a vehicle goes through a depollution process, the following vehicle's fluids are drained. This part of the depollution process involves removing all of a car's fluids liquids safely. This will include the liquids mentioned earlier and they will undertake various techniques including; filtration, distillation and reverse osmosis. Once these substances are drained the vehicle's chambers will be flushed through to ensure that no residue remains in the vehicle.

Dismantle

Once the hazardous liquids have been removed, the vehicle will be dismantled. This process involves:

Recycling Catalytic Converters

Palladium, Rhodium, and Platinum are just some of the precious metals found in catalytic converters. These metals are removed and reused in various pharmaceutical products, electronics and even jewelry – including wedding rings. Alternatively they can be reused in the production of new catalytic converters. However, regardless of what the catalytic converter is reused for, it should only ever be recycled by a trained professional as once the ceramic interior is opened and exposed to the air it is considered hazardous waste.

Recycling Car Tyres

As far as materials go tyres have always been one of the harder materials to source a second use for. If the tyres are only part worn then they can be used again on another vehicle, but if not recent developments have found interesting new uses for them. In their raw form, before they are melted down, tyres are often used as roadbeds on race tracks or running track material, or as playground mulch.

Recycling Glass

Glass is an incredibly versatile material and due to its composition can be recycled over and over again. Uses for glass can be anything from; more glass, ornaments, tiles, door-knobs or it can even be shredded down into builder's sand.

Destroy

Once the vehicle has been depolluted and dismantled it's time for destruction. Which we mean quite literally because once the vehicle has been deconstructed, the metal shell of the car will be crushed and sent to a metal mill. When a vehicle is destroyed it will go through the following process:

Magnetic Separation

Typically, the majority of steel is magnetic and so it's relatively easy to separate from other recyclable materials such as plastic.

Detinning

Most vehicles have a thin layer of tin over them which prevents the vehicle from rusting. This is great on a car however it is not necessary when a vehicle is being scrapped and recycled, and so the thin layer of tin is removed. This is done by a process called 'detinning' and happens by placing the scrap in a hot caustic soda solution which dissolves the tin coating. Once this has been done the tin can then be recovered in a variety of ways, including; evaporation and crystallisation using sodium stannate, electrolysis or using hydrous stannic oxide and acid.

Melting

Once the steel has been separated and the layer of tin removed, the steel is placed into a furnace to be melted down. Once this is done the steel will be poured into casters and rolled into new steel flat sheets. This process can be done repeatedly without the steel losing any of its strength and when the steel is repurposed into flat sheets it can reused again as many things, including; cars and construction materials such as Rebar; a material commonly used in the construction industry to strengthen and frame concrete structures.

Statistics show that this method of recycling steel uses 75% less energy and resources than producing new steel from iron ore. In fact, recycled metals actually account for around 30% of metal production worldwide! Since the year 2000 the End Of Life Vehicles Directive set and met an initial target of recycling 85% of end of life vehicles and due to its success, as of last year the target of recycling has increased to 95%. As a result of current UK legislation regarding the recycling of scrap vehicles, Authorised Treatment Facilities are required by law to make the process of recycling an end of life vehicle as simple, and as easy, as possible. Because of this a vehicle's owner is entitled to a free method of disposal when the vehicle reaches the end of its life, and as a result the owner shouldn't have to pay for the removal of their vehicle. However, there are some companies who will purchase vehicles to take to ATFs themselves, who will try and charge for the collection of a scrap vehicle.

Car Crusher

Stacks of crushed cars.

A car crusher is an industrial device used to reduce the dimensions of derelict (depreciated) cars prior to transport for recycling.

A Ford van being crushed in St. Louis MO.

A blue 1990s Lincoln Town Car after crushing.

Car crushers are compactors and can be of several types: one is a "pancake", where a scrap automobile is flattened by a huge descending hydraulically powered plate, or a baling press type, with which the automobile is compressed from several directions until it resembles a large cube. A third type is a mobile crusher. It is small enough that it can be moved from location to location with a semi truck. This machine utilizes a much smaller plate that crushes the vehicle in sections, as it is slowly fed through the machine.

Types of Car Crushers

Car Crushing Machine

One of the first car crushing machines was invented by Allen B. Sharp and Richard A. Hull, both assignors to Al-Jon Incorporated located in Ottumwa, Iowa. The patent for the machine was filed on March 22, 1965 and patented on August 16, 1966. This United States Patent was primarily examined by Walter A. Scheel.

The reason they came up with this invention is because scrap cars were too big and bulky to transport to the sites that turned them into reusable material, and the cost to transport them was unethical because, at times, it would cost more to send it than the car was worth, because transportation costs were determined by weight. Since uncrushed cars were less dense and took up more space, even for a short haul, the scrap cars were worth less than it cost to deliver them.

Before then, car crushing methods consisted of dropping heavy weights on cars, which was time consuming, sometimes costly, and produced inconsistent scrap sizes. With this car crushing machine, a car is fed through a hydraulically powered jaw and is slowly flattened as it goes through, similar to how a pasta machine flattens pasta dough. The car scraps are flattened into dimensions of six inches tall by five to six feet wide, similar to the length of its original size. The design of this machine is meant to be portable so it can move to anywhere cars have been gathered by being within the legal size of highway transport. The machine can be operated by a single person.

Mobile Car Crusher

The mobile car crusher was invented by Charlie Roy Hall in the city of Wadley, Georgia in the year 1996 and was patented on August 12, 1997. The primary examiner of this United States patent was Stephen F. Gerrity. As a condensed version of a standard car crusher, a mobile car crusher has a smaller opening with lower output, so it can crush only smaller sized, C-class cars. This machine is mobile as a result of its dual functions: a travel mode for highways, which is when the hydraulic cylinder guiding posts are lowered, and a working mode, when the hydraulic cylinder guiding posts are raised. There are two guiding posts on either side of the machine, with hydraulic cylinders inside of them. The guiding posts protect the hydraulic cylinders from external interference. The hydraulic cylinders are what is used to apply pressure to the car, with a crusher hood to spread the pressure evenly and crush the entirety of the car. A heavy-duty lowboy trailer is attached to the bottom of the mobile car crusher and can be transported only using a semi-trailer truck. Due to the increase in car production, scrap metal became high in demand, so the mobile car crusher was made to increase efficiency of gathering scrap metal. Lifespans of cars were also decreasing, so the number of cars being sent to large centralized car crushing facilities rose. Consequently, facilities needed to outsource to mobile car crushers so they could keep up with the increased volume of automobiles.

SHIP RECYCLING

Ship recycling is the complete or partial dismantling of a ship enabling the re-use of valuable materials. It is what ships face in the end of their lifespan which for the modern ships is 25-30 years.

By then, corrosion, metal fatigue and lack of parts make them uneconomical to run. The materials of the ships, especially steel, are recycled and made into new products. Any re-usable equipment, electrical devices and other items on board are also re-cycled. Even many hazardous wastes can be recycled into new products such as lead-acid batteries of electronic circuit boards. In this way, ship recycling is a notable part of the circular economy, keeping resources at use for as long as possible and minimising waste.

Ship recycling is an important industry for sustainable production and supports the developing economies of several countries. Sustainable development means meeting the needs of the present whilst ensuring future generations can meet their own needs. It has three pillars: economic, environmental and social. To achieve sustainable development in trade relations and foreign policy of the EU with third countries, EU policies in the three areas must work hand in hand together and support each other to deliver positive change.

Supporting developing nations outside Europe to follow the sustainable development goals of the EU and those defined by the UN, including important topics such as tackling migration, ensure human rights protection etc., one must start with providing safe and stable jobs locally, where needed the most. Sustaining a high rate of economic growth in such nations is therefore key.

Given the need for steel in many of the growing economies, and the EU being a net exporter of steel, ship recycling is an industry which is of great economic and social importance to a number of south Asian countries, like Bangladesh, India, Pakistan and also China.

Ship Breaking

Workers drag steel plate ashore from beached ships in Chittagong, Bangladesh.

Ship-breaking or ship demolition is a type of ship disposal involving the breaking up of ships for either a source of parts, which can be sold for re-use, or for the extraction of raw materials, chiefly scrap. It may also be known as ship dismantling, ship cracking, or ship recycling. Modern ships have a lifespan of 25 to 30 years before corrosion, metal fatigue and a lack of parts render them uneconomical to operate. Ship-breaking allows the materials from the ship, especially steel, to be recycled and made into new products. This lowers the demand for mined iron ore and reduces energy use in the steelmaking process. Fixtures and other equipment on board the vessels can also be reused. While ship-breaking is sustainable, there are concerns about the use of poorer countries without stringent environmental legislation. It is also labour-intensive, and considered one of the world's most dangerous industries.

In 2012, roughly 1,250 ocean ships were broken down, and their average age was 26 years. In 2013, the world total of demolished ships amounted to 29,052,000 tonnes, 92% of which were demolished in Asia.

The largest sources of ships are states of China, Greece and Germany respectively, although there is a greater variation in the source of carriers versus their disposal. The ship-breaking yards of India, Bangladesh, China and Pakistan employ 225,000 workers as well as providing many indirect jobs. In Bangladesh, the recycled steel covers 20% of the country's needs and in India it is almost 10%.

As an alternative to ship-breaking, ships may be sunk to create artificial reefs after legally-mandated removal of hazardous materials, or sunk in deep ocean waters. Storage is a viable temporary option, whether on land or afloat, though all ships will be eventually scrapped, sunk, or preserved for museums.

Wooden-hulled ships were simply set on fire or 'conveniently sunk'. In Tudor times, ships were also dismantled and the timber re-used. This procedure was no longer applicable with the advent of metal-hulled boats.

HMS *Queen* heeled over on the Thames foreshore off Rotherhithe.

In 1880, Denny Brothers of Dumbarton used forgings made from scrap maritime steel in their shipbuilding. Many other nations began to purchase British ships for scrap by the late 19th century, including Germany, Italy, the Netherlands and Japan. The Italian industry started in 1892, and the Japanese after an 1896 law had been passed to subsidise native shipbuilding.

After being damaged or involved in a disaster, liner operators did not want the name of the broken ship to tarnish the brand of their passenger services. The final voyage of many Victorian ships was with the final letter of their name chipped off.

In the 1930s, it became cheaper to 'beach' a boat and run her ashore as opposed to using a dry dock. The ship would have to weigh as little as possible and run ashore at full speed. Dismantling operations required a 10 feet (3.0 m) rise of tide and close proximity to a steel-works. Electric shears, a wrecking ball and oxy-acetylene torches were used. The technique of the time is almost identical to that of developing countries today. Similarly, Thos W Ward Ltd., one of the largest breakers in the United Kingdom in the 1930s, would recondition and sell all furniture and machinery. Many historical artifacts were sold at public auctions: the Cunarder RMS *Mauretania*, sold as scrap for GB£78,000, received high bids for her fittings worldwide. However, even with obsolete technology, any weapons and military information were carefully removed.

Dismantling of *Redoutable* in Toulon.

Technique

The decommissioning process is entirely different in developed countries than it is in developing countries. In both cases, ship-breakers bid for the ship, and the highest bidder wins the contract. The ship-breaker then acquires the vessel from the international broker who deals in outdated ships. The price paid is around $400 per tonne and the poorer the environmental legislation the higher the price. The purchase of water-craft makes up 69% of the income earned by the industry in Bangladesh, versus 2% for labour costs. The boat is taken to the decommissioning location either under its own power or with the use of tugs.

Developing Countries

In developing countries, ships are run ashore on gently sloping sand tidal beaches at high tide so that they can be accessed for disassembly. As described, the sizeable ship-breaking industry of Bangladesh traces its origin to a ship beached there accidentally during a cyclone. Manoeuvring a large ship onto a beach at high speed takes skill and daring even for a specialist captain, and is not always successful. Next, the anchor is dropped to steady the ship and the engine is shut down. It takes 50 labourers about three months to break down a normal-sized cargo vessel of about 40,000 tonnes.

The decommissioning begins with the draining of fuel and firefighting liquid, which is sold to the trade. Any re-usable items—wiring, furniture and machinery—are sent to local markets or the trade. Unwanted materials become inputs to their relevant waste streams. Often, in less-developed nations, these industries are no better than ship-breaking. For example, the toxic insulation is usually burnt off copper wire to access the metal. Some crude safety precautions exist—chickens are lowered into the chambers of the ship, and if the birds return alive, they are considered safe.

Gas cutting in Chittagong, Bangladesh.

Sledgehammers and oxy-acetylene gas-torches are used to cut up the steel hull. Cranes are not typically used on the ship, because of costs. Pieces of the hull simply fall off and are dragged up on the beach, possibly aided with a winch or bulldozer. These are then cut into smaller pieces away from the coast. 90% of the steel is re-rollable scrap: higher quality steel plates that are heated and reused as reinforcement bar for construction. The remainder is transported to electric arc furnaces to be melted down into ingots for re-rolling mills. In the re-rolling mills, the heating of painted steel plates (in particular, those painted with chlorinated rubber paints) generates dioxins. Substances which are costly to dispose of, such as hazardous waste, are left on the beach or set on fire, even old batteries and half-empty cans of paint. Stockpiled in Bangladesh, for example, are 79,000 tonnes of asbestos, 240,000 tonnes of PCBs and 210,000 tonnes of ozone-depleting substances (mainly chlorinated polyurethane foam).

Developed Countries

In developed countries the dismantling process should mirror the technical guidelines for the environmentally sound management of the full and partial dismantling of ships, published by the Basel Convention in 2003. Recycling rates of 98% can be achieved in these facilities. Prior to dismantling, an inventory of dangerous substances should be compiled. All hazardous materials and liquids, such as bilge water, are removed before disassembly. Holes should be bored for ventilation and all flammable vapours are extracted.

Vessels are initially taken to a dry dock or a pier, although a dry dock is considered more environmentally friendly because all spillage is contained and can easily be cleaned up. Floating is, however, cheaper than a dry dock. Storm water discharge facilities will stop an overflow of toxic liquid into the waterways. The carrier is then secured to ensure its stability. Often the propeller is removed beforehand to allow the water-craft to be moved into shallower water.

Workers must completely strip the ship down to a bare hull, with objects cut free using saws, grinders, abrasive cutting wheels, hand held shears, plasma and gas torches. Anything of value, such as spare parts and electronic equipment is sold for re-use, although labour costs mean that low value items are not economical to sell. The Basel Convention demands that all yards separate hazardous and non-hazardous waste and have appropriate storage units, and this must be done before the hull is cut up. Asbestos, found in the engine room, is isolated and stored in custom-made plastic wrapping prior to being placed in secure steel containers, which are then landfilled.

Many hazardous wastes can be recycled into new products. Examples include lead-acid batteries or electronic circuit boards. Another commonly used treatment is cement-based solidification and stabilization. Cement kilns are used because they can treat a range of hazardous wastes by improving physical characteristics and decreasing the toxicity and transmission of contaminants. A hazardous waste may also be "destroyed" by incinerating it at a high temperature; flammable wastes can sometimes be burned as energy sources. Some hazardous waste types may be eliminated using pyrolysis in a high temperature electrical arc, in inert conditions to avoid combustion. This treatment method may be preferable to high temperature incineration in some circumstances such as in the destruction of concentrated organic waste types, including PCBs, pesticides and other persistent organic pollutants. Dangerous chemicals can also be permanently stored in landfills as long as leaching is prevented.

Valuable metals, such as copper in electric cable, that are mixed with other materials may be recovered by the use of shredders and separators in the same fashion as e-waste recycling. The shredders cut the electronics into metallic and non-metallic pieces. Metals are extracted using magnetic separators, air flotation separator columns, shaker tables or eddy currents. The plastic almost always contains regulated hazardous waste (e.g., asbestos, PCBs, hydrocarbons) and cannot be melted down.

Large objects, such as engine parts, are extracted and sold as they become accessible. The hull is cut into 300 tonne sections, starting with the upper deck and working slowly downwards. While oxy-acetylene gas-torches are most commonly used, detonation charges can quickly remove large sections of the hull. These sections are transported to an electric arc furnace to be melted down into new ferrous products, though toxic paint must be stripped prior to heating.

Risks

Health Risks

70% of ships are simply run ashore in developing countries for disassembly, where (particularly in older vessels) potentially toxic materials such as asbestos, lead, polychlorinated biphenyls and heavy metals along with lax industrial safety standards pose a danger for the workers. Burns from explosions and fire, suffocation, mutilation from falling metal, cancer, and disease from toxins are regular occurrences in the industry. Asbestos was used heavily in ship construction until it was finally banned in most of the developed world in the mid-1980s. Currently, the costs associated with removing asbestos, along with the potentially expensive insurance and health risks, have meant that ship-breaking in most developed countries is no longer economically viable. Dangerous vapors and fumes from burning materials can be inhaled, and dusty asbestos-laden areas are commonplace.

Removing the metal for scrap can potentially cost more than the value of the scrap metal itself. In the developing world, however, shipyards can operate without the risk of personal injury lawsuits or workers' health claims, meaning many of these shipyards may operate with high health risks. Protective equipment is sometimes absent or inadequate. The sandy beaches cannot sufficiently support the heavy equipment, which is thus prone to collapse. Many are injured from explosions when flammable gas is not removed from fuel tanks. In Bangladesh, a local watchdog group claims that one worker dies a week and one is injured per a day on average.

The problem is caused by negligence from national governments, shipyard operators, and former ship owners disregarding the Basel Convention. According to the Institute for Global Labour and Human Rights, workers who attempt to unionize are fired and then blacklisted. The employees have no formal contract or any rights, and sleep in over-crowded hostels. The authorities produce no comprehensive injury statistics, so the problem is underestimated. Child labour is also widespread: 20% of Bangladesh's ship-breaking workforce are below 15 years of age, mainly involved in cutting with gas torches.

Several United Nations committees are increasing their coverage of ship-breakers' human rights. In 2006, the International Maritime Organisation developed legally binding global legislation which concerns vessel design, vessel recycling and the enforcement of regulation thereof and a 'Green Passport' scheme. Water-craft must have an inventory of hazardous material before they are scrapped, and the facilities must meet health & safety requirements. The International Labour Organization created a voluntary set of guidelines for occupational safety in 2003. Nevertheless, Greenpeace found that even pre-existing mandatory regulation has had little noticeable effect for labourers, due to government corruption, yard owner secrecy and a lack of interest from countries who prioritise economic growth. There are also guards who look out for any reporters. To safeguard worker health, the report recommends that developed countries create a fund to support their families, certify carriers as 'gas-free' (i.e. safe for cutting) and to remove toxic materials in appropriate facilities before export. To supplement the international treaties, organisations such as the NGO Shipbreaking Platform, the Institute for Global Labour and Human Rights and ToxicsWatch Alliance are lobbying for improvements in the industry.

Environmental Risks

In recent years, ship-breaking has become an issue of environmental concern beyond the health of the yard workers. Many ship-breaking yards operate in developing nations with lax or no environmental law, enabling large quantities of highly toxic materials to escape into the general environment and causing serious health problems among ship-breakers, the local population, and wildlife. Environmental campaign groups such as Greenpeace have made the issue a high priority for their activities.

Along the Indian subcontinent, ecologically-important mangrove forests, a valuable source of protection from tropical storms and monsoons, have been cut down to provide space for water-craft disassembly. In Bangladesh, for example, 40,000 mangrove trees were illegally chopped down in 2009. The World Bank has found that the country's beaching locations are now at risk from sea level rise. 21 fish and crustacean species have been wiped out in the country as a result of the industry as well. Lead, organotins such as tributyltin in anti-fouling paints, polychlorinated organic compounds, by-products of combustion such as polycyclic aromatic hydrocarbons, dioxins and furans are found in ships and pose a great danger to the environment.

The Basel Convention on the Control of Trans-boundary Movements of Hazardous Wastes and Their Disposal of 1989 has been ratified by 166 countries, including India and Bangladesh, and in 2004, End of Life Ships were subjected to its regulations. It aims to stop the transportation of dangerous substances to less developed countries and mandate the use of regulated facilities. However, Greenpeace reports that neither vessel exporter nor breaking countries are adhering to its policies. The organisation recommends that all parties enforce the Basel Convention in full, and hold those who break it liable. Furthermore, the decision to scrap a ship is often made in international waters, where the convention has no jurisdiction.

The Hong Kong Convention is a compromise. It allows ships to be exported for recycling, as long as various stipulations are met: All water-craft must have an inventory and every shipyard needs to publish a recycling plan to protect the environment. The Hong Kong Convention was adopted in 2009 but with few countries signing the agreement.

In March 2012 the European Commission proposed tougher regulations to ensure all parties take responsibility. Under these rules, if a vessel has a European flag, it must be disposed of in a shipyard on an EU "green list." The facilities would have to show that they are compliant, and it would be regulated internationally in order to bypass corrupt local authorities. However, there is evidence of ship owners changing the flag to evade the regulations. China's scrap industry has vehemently protested against the proposed European regulations. Although Chinese recycling businesses are less damaging than their South Asian counterparts, European and American ship-breakers comply with far more stringent legislation.

GREEN SHIP RECYCLING

The increasing waste and its improper management are one of the crises that countries across the world face these days. Be it an agricultural waste or industrial waste, the rise in the disposal of waste materials is at an alarming rate, polluting the land, air and water as never before. Studies have stated that 40 percent of the waste worldwide ends up in huge rubbish tips, and also the oceans will see more plastic in it than fish by 2050. However, the concrete efforts in the past few decades have made remarkable changes in our disposable culture and also opened doors for a number of alternatives to waste disposal. Among them, recycling has been widely accepted as one of the fruitful methods for waste management.

Like any other industry, the shipping industry, indeed world's biggest polluters, also creates a huge amount of waste every day. While ships dispose hundreds of tonnes of garbage from day to day operations, the disposing of a ship after it reaches the end of its service life also leaves a huge amount of waste, posing a potential hazard to the environment. The improper disposal of the ships in earlier days, especially when they were left unattended after discontinuation from the service, has created several graves of abandoned ships around the world. And, in the past decades, ship owners have also tried several other techniques; including Scuttling-the deliberate sinking of a ship, deep water sinking and shipbreaking, to get rid of their old vessels.

While shipbreaking has emerged as the most common method of ship disposal among them, the dirty shipbreaking practices have resulted in the dumping of dangerous toxic materials such as asbestos and PBCs on beaches and other open spaces. Sometimes, companies offload their vessels onto beaches in third world countries such as Bangladesh, India, and Pakistan, allowing locals to dismantle the vessel without taking any proper measures.

However, when recycling and re-using goods and products has become an important requirement now, the shipbreaking method has also witnessed the recycling of the parts of the vessel.

Moreover, with the rise in awareness towards the maritime environment, there have been several changes in the process, which have given rise to a new term – green ship recycling. International Maritime Organisation's Hong Kong International Convention for the Safe and Environmentally Sound Recycling of Ships, 2009 also strictly directed that vessels that are being recycled after their service lives should not pose any unnecessary risks to human health, safety and to the environment. As a viable alternative to other methods of shipbreaking that makes negative effects on the environment, green ship recycling has been introduced across the world. As a way of responsible ship recycling, this method reduces the amount of waste and also keeps the waste materials from shipbreaking out of the beaches, reducing its impact on the environment.

There are several reasons which have made the concept of green ship recycling popular and meaningful. But, the most relevant benefits among them are:

- Isolate those parts of the ship which are harmful and dangerous to both marine and human lives.

- Conserve marine ecosystem by proper discarding of ship breaking waste.

- Reusing those parts of the ship that are important and can be re-used successfully while making new ships, thus saving resources.

- Help the shipowner benefit from the process by optimum utility of the ship's parts.

The valuable components of a ship that are reused include steel, aluminium, silver and brass, among others. Since a major part of a ship's weight is in steel, the steel scrap from the vessel is being converted into bars and rods for several other uses. However, in addition to the metal that can be recycled, there are a number of the toxic components inside a vessel. These harmful substances include lead, asbestos, mercury and oil sludge etc. The inefficient shipbreaking methods, especially those carried out on beaches than the dry-dock ship recycling facilities, allow these toxic and hazardous waste to be disposed of unsafely.

However, the green ship recycling, which carries great responsibility of saving the environment, offers a better recycling standard. One of the major harmful materials that are safely disposed off with the help of green ship recycling process is asbestos. Any great informational site about asbestos will tell you that asbestos has been banned from being used in ships from the past two decades. But the ships in which asbestos had been used initially need to be recycled now. Since continuous exposure to asbestos can cause problems not just to the marine life forms but also to the people aboard the ship, this toxic component is being recycled with greater caution under this process.

Unlike the unhealthy process in which the dismantling of the vessel occurs on beaches, the green recycling centres with dry-dock facilities capture the toxic waste properly and dispose it without allowing them to flow out to waterways. Many green ship recycling labs are so well equipped that the success rate for the disposal of the harmful materials is nearly around 99%. In addition to protecting the environment, these green recycling centres are also offering more green recycling jobs, offering a safe workplace for the labourers.

On the other hand, this environmentally sound and safe recycling of a ship also offers the owner optimum utility of the ship's parts. With the methodical dismantling of the vessel, the components that can be reused are saved with better care. The steel, along with other metal components, turns into rods for use in the construction industry and also corner castings and hinges. The generators and batteries which were part of the scrapped vessel will be reused for several other purposes. The appropriate recycling of the hydrocarbons on board transformed into oil products, while light fittings also reused on another vessel or even on land.

Permissions

Index

A

Aluminum, 3, 5-6, 11, 31-32, 57, 74-77, 89-90, 96-97, 157, 176, 180-183, 186, 190, 197, 209

Appliance Recycling, 141-144

Asphaltic Material, 137

Atmospheric Distillation, 137-139

Automotive Oil Recycling, 131

B

Battery Recycling, 144-149

Blast Furnaces, 14, 132

Blasting Process, 26

C

Cardboard Recycling, 10, 128-131

Cement Manufacturing, 117, 163, 165

Chemical Waste, 37

Conserve Energy, 6, 38, 94

Cost-benefit Analysis, 27

Crude Sorting, 160

Crystalline Silicon, 178, 181-182

D

Demineralisation, 134-135

Dimethyl Carbonate, 150

Dissolution Process, 153

Downcycling, 39-41, 45, 178, 180

Drug Recycling, 171-172

E

Electrochemical Process, 154, 156

Electronic Waste, 143, 180, 188-189, 192, 195-196, 198, 201-202, 207

Energy Consumption, 20, 25, 48, 101, 108, 118, 154

Energy Savings, 11, 13, 74, 90, 101, 108

Environmental Conservation, 130

Extended Producer Responsibility, 144, 178, 202

F

Ferrous Metals, 20, 32-34, 57, 71-72, 74-78, 142, 194

Ferrous Scrap, 21, 74, 77, 141

Fluorescent Tube, 157-158

Functional Unit, 27, 30

H

Hydraulic Force, 33

Hydrometallurgical Process, 152, 154

I

Injection Molding, 7

L

Landfill, 10-14, 18, 37, 40, 43-44, 57, 75, 81, 86, 94, 102, 106, 109, 115, 117, 140, 149, 157, 160-161, 164, 166-167, 178, 180, 187, 192, 198, 202-203

Landfill Disposal, 164, 203

Lead Contamination, 145

Legal Dumping, 164

Life Cycle Costing, 27, 29

Lithium, 145-147, 149-155, 200

Loss Of Biodiversity, 6-7

M

Material Quality, 24, 46

Mature Market, 19

Medicine Recycling, 172, 174

Melt Volume Rate, 7

Metal Recycling, 32-33, 74-76, 96

Metal Scraps, 19

Methane, 37, 47, 106-107, 164

Municipal Solid Waste, 23, 38, 57, 109-111, 144, 161, 188, 199, 203

N

Non-ferrous Metals, 20, 32-34, 74-78, 194

P

Packaging Waste, 13, 36

Polyethylene Terephthalate, 8, 17, 40, 47, 66, 80-82, 162

Polyvinylidene Fluoride, 150

Product Design, 54

Product Recycling, 128-129, 131, 133, 135, 137, 139, 141, 147, 149, 155, 157, 159, 165, 167, 169, 171, 173, 175, 177, 179, 181, 183, 185, 187

Pulverized Coal, 14

Pyrometallurgical Process, 153, 156

R

Reaction Tank, 135

Recursive Recycling, 57-58

Recyclable Items, 2

Recyclate, 7-9, 70, 87-88

Recycled Paper, 3, 16, 63, 107-108, 110-111

Recycling Codes, 64, 70

Recycling Program, 2, 118-119, 148, 171-174, 177

Recycling Symbol, 62-64, 66-69, 81, 89, 95, 170

Renewable Energy, 6, 22, 178-179

Resin Identification Code, 64, 81-82, 170

Reverse Commerce, 58

S

Scrap Metal, 20, 30-33, 43, 72, 74, 96-97, 218, 223

Shredding, 24-25, 31, 79, 83, 99, 129, 142, 184

Silver Oxide, 146

Solid Waste, 12, 23, 34, 36, 38, 57, 78, 81, 96, 98, 109-111, 136, 141, 144, 161, 179, 188, 199, 202-203

Solid-state Recycling, 35

Source Reduction, 51, 56

Styrene Monomer, 68

Surplus Materials, 30

T

Textile Recycling, 158-159, 161

Tire Recycling, 163

U

Upcycling, 42-47

V

Vacuum Distillation, 132-133, 137-139

W

Waste Disposal, 1, 6, 13, 24, 37, 39, 51, 53, 116, 128, 131, 179, 188, 198, 203, 206, 208, 225

Waste Hierarchy, 13, 48-52, 119

Waste Minimization, 56, 206

Waste Polymers, 14

Wrecking Yard, 31

www.ingramcontent.com/pod-product-compliance
Lightning Source LLC
Chambersburg PA
CBHW080402190526

45161CB00003B/109